STUDIES IN IMPERIALISM

General editors: Andrew S. Thompson and Alan Lester
Founding editor: John M. MacKenzie

When the 'Studies in Imperialism' series was founded by Professor John M. MacKenzie more than thirty years ago, emphasis was laid upon the conviction that 'imperialism as a cultural phenomenon had as significant an effect on the dominant as on the subordinate societies'. With well over a hundred titles now published, this remains the prime concern of the series. Cross-disciplinary work has indeed appeared covering the full spectrum of cultural phenomena, as well as examining aspects of gender and sex, frontiers and law, science and the environment, language and literature, migration and patriotic societies, and much else. Moreover, the series has always wished to present comparative work on European and American imperialism, and particularly welcomes the submission of books in these areas. The fascination with imperialism, in all its aspects, shows no sign of abating, and this series will continue to lead the way in encouraging the widest possible range of studies in the field. 'Studies in Imperialism' is fully organic in its development, always seeking to be at the cutting edge, responding to the latest interests of scholars and the needs of this ever-expanding area of scholarship.

Science at the end of empire

Manchester University Press

SELECTED TITLES AVAILABLE IN THE SERIES

WRITING IMPERIAL HISTORIES
ed. Andrew S. Thompson

EMPIRE OF SCHOLARS
Tamson Pietsch

HISTORY, HERITAGE AND COLONIALISM
Kynan Gentry

COUNTRY HOUSES AND THE BRITISH EMPIRE
Stephanie Barczewski

THE RELIC STATE
Pamila Gupta

WE ARE NO LONGER IN FRANCE
Allison Drew

THE SUPPRESSION OF THE ATLANTIC SLAVE TRADE
ed. Robert Burroughs and Richard Huzzey

HEROIC IMPERIALISTS IN AFRICA
Berny Sèbe

Science at the end of empire

EXPERTS AND THE DEVELOPMENT OF THE
BRITISH CARIBBEAN, 1940–62

Sabine Clarke

MANCHESTER UNIVERSITY PRESS

Copyright © Sabine Clarke 2018

The right of Sabine Clarke to be identified as the author of this work has been asserted by her in accordance with the Copyright, Designs and Patents Act 1988.

An electronic version of this book is also available under a Creative Commons (CC-BY) licence, thanks to the support of the Wellcome Trust, which permits distribution, reproduction and adaptation provided the author(s) and Manchester University Press are fully cited. Details of the licence can be viewed at http://creativecommons.org/licenses/by/4.0/

Published by MANCHESTER UNIVERSITY PRESS
ALTRINCHAM STREET, MANCHESTER M1 7JA
www.manchesteruniversitypress.co.uk

British Library Cataloguing-in-Publication Data
A catalogue record for this book is available from the British Library

ISBN   978 1 5261 3138 6   hardback
ISBN   978 1 5261 3140 9   open access

First published 2018

The publisher has no responsibility for the persistence or accuracy of URLs for any external or third-party internet websites referred to in this book, and does not guarantee that any content on such websites is, or will remain, accurate or appropriate.

Typeset by Out of House Publishing
Printed by Lightning Source

For my grandfather, Leon Norton

CONTENTS

List of figures and tables—page viii
Acknowledgements—ix
Abbreviations—xi

Introduction		1
1	New uses for sugar	21
2	Scientific research and colonial development after 1940	49
3	'Men, money and advice' for Caribbean development	76
4	Laboratory science, laissez-faire economics and modernity	104
5	An industrialisation programme for Trinidad	129
6	Bringing research 'down from the skies'	154
Conclusion: Science and industrial development: lessons from Britain's imperial past		181

Bibliography—193
Index—202

FIGURES AND TABLES

Figures

1 Map of the Caribbean in 1953. (Reproduced with permission from the National Archives, Kew Gardens.) *page* xii
2 Map of Trinidad and Tobago. (Public domain, via Wikimedia Commons.) xii
3 Norman Haworth. (Public domain, via Wikimedia Commons.) 42
4 Aerial view of the Imperial College of Tropical Agriculture, c.1951. The experimental sugar factory is in the foreground. (Reprinted with permission of the University of the West Indies, St Augustine, Trinidad.) 112
5 Staff of the ICTA, c.1951. Wiggins is seated in the front row, fourth from the left 116

Tables

1 Principal exports of the British Caribbean colonies in 1947 13
2 Population density in the Caribbean territories, 1950 15
3 Research institutions in Britain's colonies funded by the CDW Acts, 1940–52 61

Every effort has been made to obtain permission to reproduce copyright material and the publisher will be pleased to be informed of any errors and omissions for correction in future editions.

ACKNOWLEDGEMENTS

This book has its origins in my time at the Centre for the History of Science, Technology and Medicine at Imperial College, London, and I am very grateful to David Edgerton for the guidance he gave me when I first began developing my ideas, and for the many times he has cast a critical eye over my arguments. I am also indebted to former colleagues at the Wellcome Unit for the History of Medicine at the University of Oxford – Sloan Mahone, Mark Harrison, Margaret Jones, Karen Browne, John Manton and Maggie Pelling. I thank them for their encouragement, advice and friendship. In the writing of this book, I have been fortunate to have the benefit of incisive feedback from David Clayton and Henrice Altink at the University of York, and my gratitude also goes to Richard Bessel, Stuart Carroll, Bill Sheils and Catriona Kennedy for conversations that helped me clarify my thoughts. In addition, I want to thank several cohorts of students who have taken my Special Subject at York; their thoughts on the histories of development we studied gave me much inspiration and enormous pleasure.

Versions of some of the chapters of this book have been presented at conferences and seminars in the Europe, the USA and UK, and the discussions that followed were helpful and stimulating. I would like to thank in particular Michael Worboys, Jonathan Harwood, David Killingray, Prakash Kumar, Desiree Schauz, Robert Bud, Mary Chamberlain, Viviane Quirke, Sally Horrocks, Paul Mosley, Fern Elsdon-Baker, Jon Agar, Brian Balmer, Jeff Hughes, Casper Andersen, Graeme Gooday, Wenzel Geissler, Rita Pemberton and Debbie McCollin.

This book would have been impossible without a fellowship from the Wellcome Trust that enabled me to carry out my research and travel to Trinidad, Barbados and the USA, and I would like to thank the Trust for its support. The success of my research trips was the result of the help and expert advice I received from the archivists at UWE, St Augustine, Trinidad and the National Archives of Trinidad and Tobago, the Barbados National Archives and the National Archives and Records Administration at College Park, Maryland, plus of course, the team at the National Archives in London. I am extremely grateful to all of them.

ACKNOWLEDGEMENTS

I would like to thank my parents, Tony Clarke and Francoise D'arcy, and Colette and David Holloway, for various kinds of help that allowed me to pursue a new career in history. Finally, I thank Tim, whose help and encouragement have been so boundless.

ABBREVIATIONS

ARC	Agricultural Research Council
BNA	Barbados National Archives
BWISA	British West Indies Sugar Association
CDC	Colonial Development Corporation
CDW Act	Colonial Development and Welfare Act
CDW Org	Colonial Development and Welfare Organisation
CEAC	Colonial Economic Advisory Committee
CMRI	Colonial Microbiological Research Institute
CPC	Colonial Products Council
CPRC	Colonial Products Research Council
CRC	Colonial Research Committee
CWT	A hundredweight
DCL	Distillers Company Ltd
DSIR	Department of Scientific and Industrial Research
EWMC	Eric Williams Memorial Collection
FRB	Fuel Research Board
ICTA	Imperial College of Tropical Agriculture
MRC	Medical Research Council
MSA	Maurice Stacey Archive
NARA	National Archives and Records Administration
NATT	National Archives of Trinidad and Tobago
PRIDCO	Puerto Rican Industrial Development Company
SCEAUCRM	Scientific Committee for Examining Alternative Uses of Colonial Raw Materials
STL	Sugar Technology Laboratory
TNA	The National Archives, London

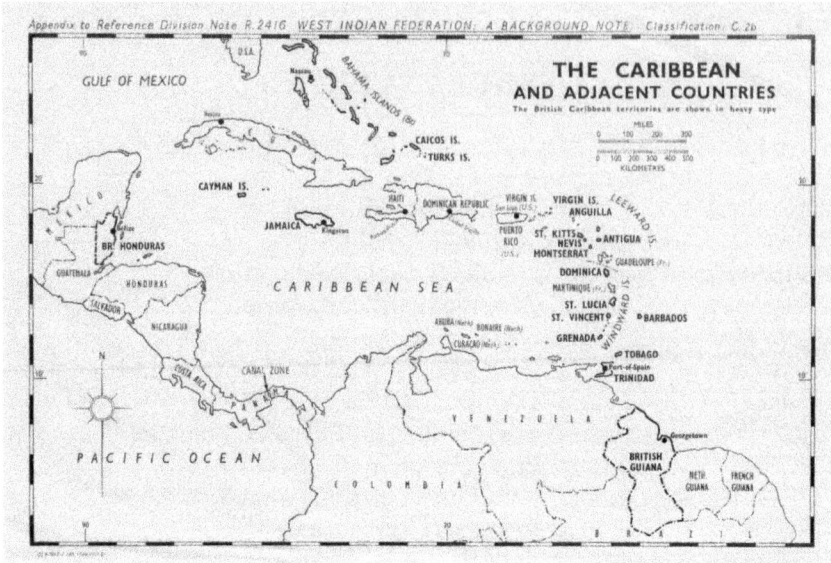

Figure 1 Map of the Caribbean in 1953.

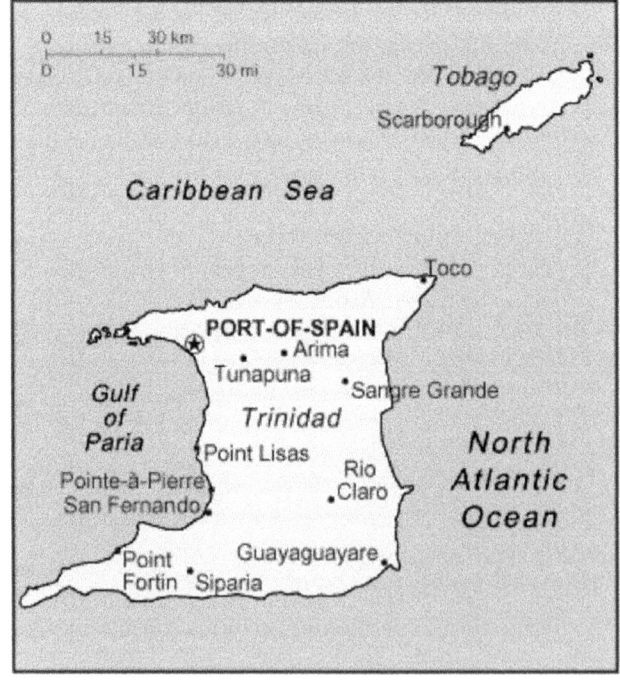

Figure 2 Map of Trinidad and Tobago.

INTRODUCTION

During the 1930s, episodes of violent protest by the inhabitants of Britain's Caribbean colonies brought the extremely poor living and working conditions that existed in these territories to domestic and international attention. Revelations of widespread unemployment, squalid housing and malnutrition threatened the moral authority of British rule and provided fuel for critics of British imperialism. As a result, Britain made a commitment to improving living conditions in an area of the British Empire that it had previously neglected. This book explores the function of knowledge and expertise in the visions of economic development that were subsequently produced for the region, with a focus on the debate about encouraging new industry. Historians have often said that Britain was unwilling to sanction the growth of manufacturing in the Colonial Empire in order to protect markets for British industrial exports.[1] In fact, officials in London saw the development of secondary manufacturing in the Caribbean as essential after the Great Depression in order to raise living standards and contain political dissent. Colonial Office plans included a vision of economic development that gave a key role to scientific research. The Colonial Office was inspired by recent discoveries such as nylon, polythene and penicillin to sponsor laboratory research that would transform sugar from a low-value foodstuff into a starting compound for the expanding chemical and fermentation industries. The expectation was that new factories producing the constituents of plastics, drugs and fuels from sugar would be established in the Caribbean itself. In this vision of industrialisation, state-funded research would enable Britain's Caribbean colonies to participate in the emerging 'brave new synthetic world', and in doing so these places would find their economic fortunes revived.[2]

By exploring post-war visions of economic development for the British Caribbean colonies this work produces a rethinking of our wider understanding of the history of science and development in the twentieth century. Despite the rise of development as a universal ideal for the Global South and the emergence of development studies as a major scholarly field, we employ a narrative of past projects that can be partial and even erroneous in its claims. By the 1950s, the priority for most governments, academics and agencies concerned with the advancement of low-income countries was identifying the necessary incentives for industrialisation. In exploring the inspiration for such measures, scholars have focused on models provided by economists such as W. Arthur Lewis, Raul Prebisch and W. W. Rostow.[3] In contrast, this account shows how ideas about industrial development were worked out in a period before the advent of famous theoretical interventions such as modernisation theory. It describes how the late colonial Caribbean was a laboratory for the emergence of new ideas about the development of manufacturing and shows how initiatives on the ground could in fact contribute to later theoretical work; a rather different relationship between theory and practice from that typically described.

This account also broadens our understanding of development by focusing on a region that has been overlooked in historical studies. The riots in Britain's Caribbean colonies during the 1930s persuaded the British government to greatly increase development spending across the Colonial Empire after 1940 in an attempt to improve conditions and mollify critics of British imperialism.[4] The focus in scholarship that explores the results of this turn to development has been on Africa, however, so that we know very little of the plans formulated for Britain's Caribbean colonies after 1940, despite the significance of the region for producing new policy in the first place. Importantly, this exploration of Britain's economic plans for the West Indies shows them to be of a very different character from the state-centred, rational and authoritarian agricultural schemes in Africa that historians have often presented as typical of development projects in the post-war period.[5] In British development plans for the Caribbean the political utility of science and expert action stemmed from their capacity to reconcile new development ambitions with long-standing laissez-faire principles favoured by the Colonial Office. In the case of research into sugar, laboratory investigations represented a mode of state intervention that struck an acceptable balance between government action and private interests. The argument that is made here is that while the late colonial period was a high-water mark for state-led development in Britain's colonies, a time when there was unprecedented emphasis on science and much talk of planning, the vision of British West Indian economic development employed by the Colonial Office was essentially liberal in character.

Aside from providing a resolution to a central issue in the political economy of industrial development, knowledge and expert advice also became increasingly important to the maintenance of Britain's control over its colonial possessions after 1945. The exceptional level of funding for scientific research made available through the 1940 Colonial Development and Welfare (CDW) Act and its successors was an important gesture to ward off domestic and international criticism of Britain's management of its colonies after the disturbances of the 1930s. In-depth research to establish the basic facts about tropical locations was said to ensure that Britain's development interventions were effective in the future, thus helping to restore the credibility of British imperial rule. Apart from this, the forty new laboratories created across the Colonial Empire after the Second World War were said to endow Britain's colonies with the ability to participate in the international advance of science, and therefore to operate as modern states. Science and scientists took on unprecedented importance for the British Colonial Office after 1940, both in providing solutions to practical issues that arose from the drive for development and also in demonstrating the enduring value of British interventions in the tropics and Britain's commitment to modernising its colonial possessions.

The provision of knowledge and advice was also an important strategy for the maintenance of British influence over Caribbean affairs at a time when these colonies were undergoing a process of constitutional reform that appointed local politicians to colonial legislatures in greater numbers. While the Colonial Office had clear preferences with regard to the way industrialisation should proceed in the British West Indies, responsibility for working out the details of policies for industrial development did not reside with officials in London but instead lay with the increasingly autonomous governments of the region. With the decline of direct modes of metropolitan control, Britain came to rely on the activities of scientists and expert advisors to maintain its standing with its Caribbean colonies. The idea that development projects had a key role in demonstrating the benefits of continued association with Britain was challenged, however, by the desire of the USA for a larger presence in the Caribbean area. The promotion of industrial development became part of America's strategy to promote its economic and strategic interests in the region. Britain's authority was threatened by an influx of experts disseminating development ideas that diverged significantly from those of the Colonial Office.

The co-existence of different models for development in the late colonial period helps problematise the notion that, when it came to the plans of governments and new inter-governmental agencies such as the World Health Organization, one particular paradigm of development was dominant after 1945.[6] By broadening the scope of analysis beyond agriculture to include industry, and considering Caribbean as well as African locations, we find that rather than comprising a hegemonic set of discourses and practices that privileged planning and state intervention, there was diversity in development visions in the 1940s and 1950s, and Caribbean political leaders were presented with initiatives to promote economic development that varied considerably according to their source. One aim here is to disturb the universalising tendency in some accounts of development in the past by paying far greater attention to the relationship between official development visions and the economic and political priorities of different groups of policy makers. In other words, this work aims to put politics back into our understanding of development after 1940 by showing how state-produced definitions and visions of development could contain expressions of very different roles for government and for science, and how these proposals were contingent upon the wider political and economic beliefs, ambitions and needs of those that hoped to shape the development of the Caribbean and ensure their place in its future.

Science, development and empire

It is not uncommon for scholars to claim the birth of development occurred in the post-war period, often marked by President Truman's Inaugural Address of 1949.[7] This assertion can come as a surprise to historians who have looked at the increasing focus on development in the European empires from the late nineteenth century. Historians of British imperialism have described how science was firmly implicated in the rise of colonial development as a goal of imperial policy, beginning around 1895, when Joseph Chamberlain was appointed Secretary of State for the Colonies. Michael Worboys and Joseph Hodge in particular have shown that the reformulation of 'improvement' as the goal of imperial policy to 'development' was marked by new state provision for transport, communications and science in the colonies, promoted as necessary measures to encourage business in the fullest exploitation of Britain's tropical possessions.[8] Scientific interventions were expected to help unlock the unrealised potential of the tropics. In the early twentieth century, new agricultural departments were created across the Colonial Empire and institutions for tropical medicine, entomology and the assessment of tropical products were established in Britain by the outbreak of the First World War. Funds for initiatives in health and agriculture before 1940 came from a series of acts, including the 1929 Colonial Development Act that had the wider purpose of alleviating unemployment in Britain by stimulating demand for capital goods in the colonies.[9] The drive to increase agricultural efficiency and open up territories for greater exploitation led to an expansion in scientific and medical officers in Britain's colonies, and these officers worked to exert control over tropical environments and subject populations, through projects to control tsetse and sleeping sickness, for example.[10] By the interwar period, the belief had arisen that African peoples mismanaged their land, and so experts were deployed not just to increase outputs but also to protect the environment from the apparent pressures of over-cultivation.[11]

The British government made its most concerted attempt to develop its tropical possessions after 1940. This late colonial push for development occurred alongside American and United Nations technical assistance programmes. Little work considers the interchanges that occurred between different governments and institutions with development ambitions.[12] Not only, then, does the concept of development have a history that pre-dates the Second World War, the period between 1940 and 1965 saw a more complex array of imperial, trans-imperial and international formations orientated towards development than many accounts describe.[13] This book moves from an investigation of the Colonial Office vision of Caribbean industrial development, with its focus on scientific research, to examination of debates that occurred at meetings of the US-led Caribbean Commission and in the British colony of Trinidad. It examines how ideas emerged, and circulated, at the metropolitan, regional and colonial level, and explores the significance of interactions between British and US officials and Caribbean intellectuals and politicians in shaping development thinking and practices. In doing so, it shows how an influx of foreign expertise that promoted a Caribbean-wide approach to development threatened imperial integrity in the late colonial Caribbean and disturbed British expectations of a close relationship with the British West Indies after independence.

The account of British economic development plans for the Caribbean presented here reveals a conception of state-conceived development that does not conform to the picture often presented in the literature. Historical studies of development after 1945 have often drawn on the work of James Scott and James Ferguson with a focus on large-scale agricultural projects that aimed to transform rural Africa.[14] In this scholarship, rural development schemes are described as comprising a regime. Development represents an all-encompassing form of state power – authoritarian, intrusive and dealing in standardised and regimented units of production. For many scholars, the exemplar of the development project of the twentieth century is the large-scale African agrarian scheme in which communities were uprooted, resettled in new villages and made to work on uniform plots under close supervision, as in the Gezira cotton-growing project in British-controlled Sudan, the Sukumaland Scheme in Tanganyika, or, for some historians, the Groundnut Scheme.[15] A number of accounts have sought to show how attempts to control subject populations through development projects were often limited in practice by the incompetence of officials and shortages of personnel and funding.[16] In addition, colonised people resisted official development visions on the ground.[17] What is missing, however, is the notion that governments could promote other modes of development apart from those that relied upon centralised state direction and a willingness to intervene in people's lives in order to completely remake existing patterns of living and working.

In contrast to this prevailing picture of state-led development, this book shows that the British vision of industrial development for the Caribbean was cautious about affording too large a role to the state, was financially conservative and embodied a preference for change that officials described as 'naturally occurring'. Scientific research, specifically 'fundamental research', had a function in producing interventions that conformed to this laissez-faire vision of Caribbean development. The fact that the Colonial Office favoured a rather different approach to encouraging the industrial development of the Caribbean in comparison to agricultural development in rural Africa is hardly surprising. In contrast to the African colonies, which were spoken of in terms of unexploited potential, the British West Indies were perceived as a region in decline, plagued by the problems of modern industrial society – slums, unemployment and labour unrest. Africa was a place where agricultural production was often still the mainstay of colonial economies and, in addition, increases in outputs of tropical products were expected to make a contribution to solving Britain's own economic problems after 1947. The creation of new industry in the Caribbean did not have the same economic value for Britain as increases in dollar-earning agricultural products and therefore it was not promoted through intensive and intrusive models of development. The intention here, however, is to raise questions about what we consider to be typical of development in the twentieth century and to suggest that state-produced development visions were more varied and contingent than our existing literature allows.

In exploring the relationship between science and industrial development, this book focuses on one particular group of scientific

INTRODUCTION

practitioners: research scientists who worked in a British or colonial laboratory. The aim is to explore the importance attached to scientific researchers, as a sub-group of experts, for Britain's development effort after 1940. The CDW Acts privileged scientific research, providing a substantial fund of £1 million each year from 1945 entirely for the promotion of this activity across the Colonial Empire. To put this into context, the Research Fund of the CDW Acts made the Colonial Office the second-largest funder of civil scientific research in Britain during the 1940s, with more money than the Medical Research Council (MRC) or Agricultural Research Council (ARC). The result of this focus on promoting research was the creation of over forty colonial laboratories and research institutions in Britain's colonies. Scientific advisors to the Colonial Office expended a great deal of effort differentiating between the in-depth study of colonial problems that would occur in these new centres and other types of technical work such as preparing vaccines, clearing bush to tackle tsetse fly or directing African farmers to plant new crops. The distinction between 'research', often 'fundamental research', and 'problem solving', 'routine tasks' or 'extension work' was extremely important. The claim that was made was that highly qualified scientific researchers, those that would normally work in a British university or research unit, would only work in the colonial services if they were given assurance that they would have the freedom to choose their own research problems comparable to the freedom they were said to enjoy at home. Arrangements for colonial research, therefore, needed to ensure autonomy for scientists, and it was accepted at the Colonial Office that this condition was key to the professional status of scientific researchers. The notion that research required particular working conditions, different from those of other grades of scientific or medical staff, provided a rationale for the particular apparatus introduced for colonial research across the Colonial Empire after 1940.

Recognising that emphasis on research led to changes in the arrangements for colonial science prompts a rethinking of the story of science and development in the mid twentieth century. This book is a study of the relationship between scientific research and colonial development that pays close attention to the distinct position of scientific researchers with respect to the technical services and colonial administrations, and in doing so attempts to problematise our understandings of expertise in a new way. The increasing authority claimed by experts working in the colonies in the first half of the twentieth century has been the subject of a great deal of critical comment. Scholars have spoken of a belief in the innate superiority of Western science that led to the marginalisation of local knowledge, and the imposition of unsuitable schemes on communities who did not want them but had to suffer the social and ecological consequences.[18] More recent work has shown that expert understandings of African environments and societies could in fact vary a great deal.[19] Some officers deployed in Africa took a keen interest in indigenous farming practices, for example.[20] Helen Tilley, Peder Anker and Joseph Hodge have shown that concern about the failure of previous development initiatives, interest in tropical soils, and the rise of ecology and anthropology contributed to increasingly sophisticated understandings of African environments during the interwar period.[21] Suzanne Moon and Donna Mehos have argued that we should not conflate the itinerant consultants working for international development institutions in the post-independence era with the scientific staff of the European empires that had spent lengthy periods in one location and gained much place-based knowledge.[22]

Despite the production of increasingly nuanced accounts of expertise, existing scholarship has not paid much attention to the distinctions that existed between roles in the technical services in the British colonies. This is not an argument for seeing the diversity of views held by scientific officers in the colonies but instead for greater engagement with the fact that technical officers were members of highly stratified services in which distinctions on the basis of qualifications and professional status were very important. Different grades of officer were involved in very different types of interactions with colonial peoples and the colonial state. Non-specialist members of the Colonial Agricultural Service, for example, often with basic and general qualifications from agricultural colleges in Britain, were typically at the forefront of executing large-scale development projects that involved soil terracing or the adoption of particular methods of animal husbandry. In contrast, laboratory researchers who had specialist degrees in chemistry from places like Cambridge University studied colonial products in research institutes where they had very little regular contact with local people apart from those they employed. These researchers did not spend so much time out in the field, and were not routinely engaged in the direct manipulation of the economic and social practices of local communities in the name of improvement. Instead their work contributed to the exercise of colonial power through the production of representations of the tropics and colonised peoples; representations that were often informed by the economic priorities and racial prejudices of the imperial/colonial state. Recognising the differences that existed between research staff and extension officers in terms of their relationships with both the metropolis and colonial peoples suggests that we need to move beyond talk of colonial science or a single science–development relationship to see that science could function as part of the colonial project in various ways. Apart from anything else, this raises questions about the circulation of knowledge at the level of the colony. It is not at all clear how ideas generated by elite researchers in colonial institutions were conveyed to technical staff in the field when these individuals did not attend the same meetings, read the same journals, or even exchange annual reports.[23] The focus in this book is on a group of organic chemists and microbiologists who studied tropical products in British and colonial laboratories, with funds from the CDW Acts. These individuals were highly qualified research scientists of a type who had not been employed in the Colonial Empire in great numbers before 1940. The aim here is to establish the specific place of laboratory investigations prosecuted by these individuals in the project of colonial development after 1940, or in other words, to elaborate a research–development relationship that should not be conflated with the relationships that scholars have mapped out for other groups of technical officers.

The other argument about science and empire that runs through this book is that we can only understand the function of scientific research in the late colonial period if we pay serious attention to earlier developments in science in Britain. The architects of arrangements in the colonies were explicit that their goal was the extension of the system of research that had emerged at home to the empire as whole. Most important were the arrangements for scientific research that operated in Britain through the research council system that included the Department of Scientific and Industrial Research (DSIR), the MRC and the ARC. The research council system provided both a template for research arrangements for the colonies and also a discourse on the nature of research itself. There was little specifically colonial about the conception of research that was promoted for Britain's colonies. The obstacles to good research were presented as much the same whether an institute was in Trinidad or East Anglia. The chief goal was to ensure research workers were placed under scientific oversight to ensure they had the necessary freedom to pursue their own investigations. By removing close supervision by non-scientists or less-qualified technical staff, this arrangement worked to assure the status of the research councils as the true arbiters of research funded by the British state.

This exploration of the apparatus introduced to cultivate colonial research pays particular attention to the rhetorical value of the term 'fundamental research' in shaping new arrangements for research in Britain's colonies. Work on the research councils in Britain has sometimes assumed that when these bodies referred to their preference for 'fundamental research', this can be read as a synonym for 'pure science', a term that usually referred to investigations that were driven by the curiosity of the researcher without thought of application.[24] A tendency to conflate 'fundamental research' with 'pure science' has obscured rather than illuminated important features of research council rhetoric, however, both in relation to the domestic ambitions of these bodies and to their work at the Colonial Office. It does not help us, for example, answer the question as to why officials at the Colonial Office were persuaded that Britain's colonies needed an expansion of 'fundamental research' after 1940 if this only served to remove their control and place research at some distance from practice. The answer to the question of why officials were happy to endorse an emphasis on 'fundamental research' lies in the fact that this term was one with flexibility of meaning, encompassing more than just the notion of science for its own sake. For officials at the Colonial Office, the promotion of scientific research had both utility and symbolic power, endowing their work with greater credibility and reflecting a new conception of the scope and purpose of imperial action. For the scientists who advised the office, particular characterisations of research and research workers allowed the introduction of their preferred working conditions. The multiple connotations of 'fundamental research' were important in building a consensus between scientific advisors and officials about the direction of policy during the 1940s. This consensus began to break down, however, in the 1950s, and this book will end with a consideration of the reasons for this eventual fracturing of that earlier vision of the relationship between scientific research and colonial development.

From policy-making to practice

This account considers some of the outcomes for the Caribbean of the substantial increase in funding for development and scientific research that occurred with the passing of the CDW Act of 1940. Historians have focused on the novelty of the CDW Acts in providing grants rather than loans for the development of the colonies and a commitment to programmes concerned with social improvement for the first time.[25] This assumption of greater responsibility for providing schools, hospitals and sanitation schemes is said to have stemmed from a need to demonstrate Britain's commitment to the principle of trusteeship in order to counter both Germany's demands for the restitution of its former colonies and hostile anti-imperial rhetoric from the USA after revelations of high levels of deprivation in Britain's possessions during the 1930s.

The period after 1940 was not, of course, one of imperial expansion. By 1948, Britain had relinquished formal control in India,

INTRODUCTION

Ceylon and Palestine. India and the Dominions were dealt with by the Commonwealth Relations Office from 1947 and did not receive any part of the new CDW allocation. The loss of some former territories did not represent an overall trend of gradually diminishing British influence in the years after the Second World War, however. In 1948, the Colonial Empire consisted of a variety of colonies, mandates and protectorates. The British colonies in Africa were divided into three main areas: West Africa included the colonies of Nigeria, Sierra Leone, the Gambia and the Gold Coast; East Africa included Kenya, Uganda, Tanganyika and Zanzibar; and Central Africa was made up of Northern Rhodesia and Nyasaland. The remaining British territories in Africa were Somaliland, Bechuanaland, Swaziland and Basutoland (the white settlers of Southern Rhodesia had declared their independence from Britain in 1923). The British West Indies comprised the Bahamas, Barbados, British Guiana, British Honduras, Jamaica, Trinidad and Tobago, the Leeward Islands of Antigua, Montserrat, St Christopher, Nevis, the Virgin Islands, and the Windward Islands of Dominica, Grenada, St Lucia and St Vincent.[26] The net effect of the new development ethos that emerged in the 1940s, coupled with policies intended to cultivate political change, was to give officials at the Colonial Office a reinvigorated sense of purpose.[27] The objective was the maintenance of the remaining Colonial Empire rather than its dissolution. In the years immediately after the passing of the 1940 CDW Act, development was presented as a progressive measure, one that did not prioritise metropolitan needs over colonial ones. A second definition came to the fore after Britain's economic crisis of 1947, however, in which development meant a focus on increasing colonial production, and British possessions were once again expected to work for the benefit of Britain in the first instance. British colonial policy and practice after 1940 was therefore a combination of idealism and exploitation.

In the wake of the 1940 CDW Act, the Colonial Office made a commitment to the development of secondary industry in the Caribbean. L. J. Butler remains the only historian to have considered in any detail the Colonial Office's policy for industrialisation.[28] His focus was on the drive for import-substitution industries in West Africa and no account exists of plans for the British Caribbean. Indeed, historians of the Caribbean generally deny that Britain ever had such a vision. The existing literature tells us that when Trinidad and Jamaica created legislation to encourage industrial development after 1949, this was the result of an intervention made by the famous St Lucian economist W. A. Lewis.[29] That story does not stand up well to scrutiny, and this account shows that in the case of Trinidad, an economic advisor from London was key in shaping the new legislation so that in its first incarnation, the incentives provided for industry were closer to the preferences of the Colonial Office.

Britain's Caribbean colonies were some of the oldest territories of the Colonial Empire, with Barbados coming into British possession in 1627. Barbados was the only colony in the region to be entirely in British hands throughout the colonial period. The others variously passed through French, Dutch and Spanish control before the Napoleonic Wars. Trinidad became a British colony in 1802, and like many of the other Caribbean territories, the legacy of French and Spanish influence is apparent in the language and culture of the people. The populations of the Caribbean are diverse; along with the large numbers of people who trace their origins to Africa, a system of indentureship brought workers from East India to Trinidad and British Guiana in the nineteenth century, and communities exist across the region from places as far apart as China, Syria and South Africa. In terms of economics, until the end of the eighteenth century, a Caribbean sugar industry dependent on the labour of African slaves provided many Britons with the opportunity for profit. By the twentieth century, however, the problems of British West Indies seemed intractable. The legacy of the plantation system included an over-reliance on sugar as an export, a shortage of land for independent farmers and local food production, and limited investment in new industries. Caribbean economies were reliant on a very narrow base of exports, a problem exacerbated by economic depression and war (Table 1 shows the main exports of the British West Indian colonies in 1947). Population densities were very high in some islands such as Barbados (see Table 2), contributing to an acute shortage of adequate employment and also, in some cases, a shortage of land for peasant proprietors. Many people sought work elsewhere when times were hard, either moving to another island, such as Trinidad, where a workforce was needed for the oil industry from the early twentieth century, or further afield to the USA and Central America to build the Panama Canal.

During the Great Depression, prices for primary commodities went into steep decline and the structural problems of the British West Indies were fully exposed. Workers could not find sufficient income in struggling agricultural industries to cover the high costs of food imports, and the Caribbean colonies did not produce enough food to feed themselves. Government finances worsened with the declining value of exports and rising costs of essential imports such as rice and flour so that at the moment when Caribbean peoples were

INTRODUCTION

Table 1 Principal exports of the British Caribbean colonies in 1947 (*British Dependencies in the Caribbean and North Atlantic, 1939–1952*, Cmd 8575).

Exports	Quantity	Value (£000 sterling)
Antigua		
Sugar	18,000 tons	419
Cotton	84,000 lbs	9
Bahamas		
Craw fish	415 tons	58
Tomatoes	53,000 bushels	46
Salt	2,033,000 bushels	66
Barbados		
Sugar	82,461 tons	1,879
Molasses	7,887,000 gallons	1,147
Rum	1,462,000 gallons	9
British Guiana		
Rice	19,625 tons	478
Bauxite	1,290,000 tons	1,402
Sugar	185,000 tons	3,974
British Honduras		
Gum, chicle	634 tons	328
Wood and timber	1,041,000 cubic foot	538
Grapefruit juice	2,652 tons	104
Dominica		
Lime juice	347,000 gallons	37
Essential oils	54,000 gallons	39
Cocoa	210 tons	33
Grenada		
Nutmegs	1,770 tons	442
Mace	295 tons	156
Cocoa	2,311 tons	339
Jamaica		
Rum	2,306,000 gallons	2,570
Sugar	128,000 tons	2,656
Bananas	5,520,000 stems	2,049
St Christopher-Nevis		
Sugar	32,000 tons	756
Salt	7,319,000 tons	21
Cotton	462,000 lb	48

(*continued*)

Table 1 (Cont.)

Exports	Quantity	Value (£000 sterling)
St Lucia		
Sugar	3,941 tons	88
Cocoa	416 tons	66
St Vincent		
Arrowroot	2,976 tons	86
Copra	1,185 tons	45
Cotton	229, 000 lb	26
Trinidad and Tobago		
Cocoa	4,022 tons	668
Sugar	90,000 tons	1,690
Petroleum	Crude – 31 mill. gallons	13,694
	Refined – 772 mill. gallons	

most in need of help, their governments were unable or unwilling to provide it. The Caribbean colonies were neglected by the metropolis, their issues only attracting concerted attention in the wake of protests and rioting as occurred on an unprecedented scale during the 1930s. The unrest of the interwar period combined grievances about economic privation with demands for political reform and independence.[30]

In the period after 1940, constitutional reforms were introduced across the British West Indies. Jamaica was the first British Caribbean colony to attain universal adult suffrage in 1944. Elections with full suffrage were held in Trinidad in 1946 and a ministerial system was introduced in 1950 with key roles taken by Trinidadian politicians. The pace of this type of reform varied from colony to colony in the Caribbean and the situation was made more complex by the debate around the creation of a West Indies Federation as a political and economic union of the British colonies of the region. From the perspective of Colonial Office officials, the challenge was to steer increasingly autonomous legislatures, populated by politicians who could be suspicious of metropolitan priorities, to follow the policy lines that they favoured. As Caribbean territories moved towards independence and America sought to shape the future of the region, the provision of scientific and economic advice became a key strategy for the maintenance of British power.

INTRODUCTION

Table 2 Population density in the Caribbean territories, 1950 (*British Dependencies in the Caribbean and North Atlantic, 1939–1952*, Cmd 8575).

Territory	Area (square miles)	Estimated population in mid 1950	Density (people per square mile)
Bahamas	4,404	79,000	18
Barbados	166	209,000	1259
Bermuda	21	37,000	1761
British Guiana	83,000	420,000	5
British Honduras	8,867	67,000	8
Jamaica	4,411	1,403,000	318
Leeward Islands:			
Antigua	171	45,000	263
St Christopher-Nevis and Anguilla	153	48,000	314
Montserrat	32	13,500	422
Virgin Islands	67	6,500	97
Trinidad and Tobago	1,980	627,000	317
Windward Islands:			
Dominica	305	54,000	177
Grenada and Carriacou	133	77,000	579
St Lucia	233	79,000	339
St Vincent and the Grenadines	150	67,000	447

This book moves from an investigation of the Colonial Office vision of Caribbean industrial development, with its focus on laboratory research, to examination of debates about the appropriate road to industrialisation that occurred at meetings of a new regional body, the US-led Caribbean Commission, and then at the level of the colony. The book begins with a description of the conditions during the Great Depression that existed in Britain's West Indian colonies that prompted widespread protest, before exploring debate amongst British officials, scientists and economists about the best way to address Caribbean economic problems. It shows that officials in London contrived a solution that diverged significantly from that envisaged in the famous Moyne Report. On the assumption that the Great Depression had shown the world market for sugar as a foodstuff to be saturated, the Colonial Office decided to made plans to transform sugar into a raw material to make fuels and chemical products. This vision was inspired by the rapid growth of the synthetic chemical industry in Britain, Germany and the USA in the interwar period that produced an expanding range of plastics, medical products and dyes. By 1942 the Colonial Office had created a new body, the Colonial Products Research Council (CPRC), to sponsor scientific research into finding new industrial uses for sugar and other tropical products. Chapter 2 examines the relationship between scientific investigation and colonial development that was embodied in the arrangements that emerged for colonial research during the first half of the 1940s and shows the important rhetorical and symbolic functions of scientific research for British colonial policy-making after 1940.

Chapter 3 describes the plans for colonial industrialisation that were formulated by the Colonial Office in the 1940s before placing these in the context of wider debates about economic diversification. The assumption by the Colonial Office that the colonies would follow its advice when it came to encouraging new industry was disturbed by the creation of the Caribbean Commission. This body had the ostensible role of coordinating policy for the Caribbean between the US and Britain in the first instance, but in reality it operated as a vehicle that allowed the US to expand its influence in the region. The Commission was a problem for Britain as it promoted a model of development that gave a far bigger role to the state in planning, funding and facilitating the growth of new industry than the Colonial Office deemed prudent. The contribution of this book is to show how debate at meetings of the Caribbean Commission about industrialisation in the region was a key area where wider British and US political and economic aspirations for the post-war world came into conflict.[31]

Chapters 4 and 5 look in detail at the results of new commitments to scientific research and Caribbean industrialisation at the level of the colony, in this case, the islands of Trinidad and Tobago (referred to as 'Trinidad' in this book). Chapter 4 explores the origin and significance of two new laboratories in Trinidad that undertook research in sugar chemistry and microbiology with the goal of encouraging new chemical industry. The debates of the 1940s on the best way to foster economic diversification discussed in Chapter 3 revealed a tendency amongst British officials to discourage the adoption of measures that were considered too state-centred and protectionist in nature. Funding scientific research to identify industrial uses for sugar, however, represented a resolution of the issue of how to take some action to encourage industry whilst still adhering to laissez-faire principles. The two laboratories created in Trinidad were described as places of fundamental research, meaning research into widely occurring, general phenomena, and this designation worked to carefully distinguish actions undertaken by the state in the name of development from

INTRODUCTION

more narrow activities that were considered to be rightly the business of the firm.

Chapter 5 reconstructs the process by which legislation to encourage industry was passed by the increasingly autonomous government in Trinidad. It provides an important re-evaluation of the story that is typically told about negative British attitudes towards Caribbean industrialisation and the crucial role played by Lewis in the genesis of legislation in the region. Despite the threat presented by alternative models of development, including those promoted by the US, the Colonial Office was initially successful in steering policy for industry in Trinidad along lines it saw as desirable until the 1956 elections that brought Eric Williams to power. This success was achieved not by direct instruction by London but through the judicious use of expert advisors who promoted the more liberal road to development favoured by the Colonial Office.

The final chapter examines the outcomes of the scheme to foster new industry through scientific research into new uses for sugar. By the early 1950s, officials at the Colonial Office were concerned that the work overseen by the CPRC was not making a tangible contribution to the economic development of the colonies, and the Colonial Office reorganised research in Britain and Trinidad so there was less focus on long-term fundamental research. The early 1950s saw a significant change of heart at the Colonial Office and this chapter considers the external and internal factors that contributed to the demise of the consensus that had previously existed that undirected fundamental research had an important role to play in economic development.

Science at the End of Empire shows the importance of expert advisors in attempts to influence the direction of industrial development in the Caribbean and the ways in which competition between the US and UK was played out through the politics of expertise. It demonstrates how scientific and economic advice enabled the Colonial Office to maintain political authority and influence at a time when Britain's ability to ensure a continuing relationship with its increasingly autonomous colonies was made difficult by poor economic conditions at home and the new role that America had assumed on the world stage. We can also see something of the ways in which political conditions and aspirations at the level of the colony and the region informed the responses of Caribbean legislators to the very different visions of industrialisation that were promoted after 1940. Finally, we can see how the rapidly changing political and economic conditions of the post-war period determined the success or failure of the various initiatives conceived to help the British West Indies see their fortunes transformed, including the hope that cane sugar could be reinvented as an industrial raw material.

Notes

1. M. Harrison, *Jamaica, the Caribbean and the World Sugar Industry* (New York: New York University Press, 2001); R. Kiely, *The Politics of Labour and Development in Trinidad* (Kingston: The University of the West Indies Press, 1996), pp. 5–6; B. datt Tewarie and R. Hosein, *Trade Investment and Development in the Contemporary Caribbean* (Kingston: Ian Randle, 2007); A. Payne and P. Sutton, *Charting Caribbean Development* (London: Macmillan Caribbean, 2001), pp. 2–3.
2. *The Manchester Guardian*, "Chemicals in war", 25 May 1945, p. 4.
3. Kiely, *Politics of Labour*; R. Kiely, *Industrialisation and Development: A Comparative Analysis* (London: UCL Press, 1998); H. W. Arndt, *Economic Development: The History of an Idea* (Chicago: The University of Chicago Press, 1987).
4. S. Constantine, *The Making of British Colonial Development Policy* (London: Maurice Temple Smith, 1984), ch. 7; D. J. Morgan, *The Official History of Colonial Development* (London: Macmillan, 1980), vol. 1, ch. 4; D. Goldsworthy, *Colonial Issues in British Politics, 1945–1961* (Oxford: Oxford University Press, 1971), p. 11; M. Havinden and D. Meredith, *Colonialism and Development: Britain and Its Tropical Colonies, 1850–1960* (London and New York: Routledge, 1993), pp. 199–205; J. M. Lee and M. Petter, *The Colonial Office, War and Development Policy: Organisation and the Planning of a Metropolitan Initiative, 1939–1945* (London: Maurice Temple Smith, 1982); S. R. Ashton and S. Stockwell, *Imperial Policy and Colonial Practice, 1925–1945*, British Documents on the End of Empire (London: HMSO, 1996); and the work of Caribbean scholars such as B. Brereton, *A History of Modern Trinidad, 1783–1962* (London: Heinemann, 1983).
5. The dominance of one archetype can be attributed to the enormous impact of J. Scott, *Seeing Like a State: How Certain Schemes to Improve the Human Condition Have Failed* (Yale: Yale University Press, 1998).
6. F. Cooper and R. Packard (eds), *International Development and the Social Sciences: Essays on the History and Politics of Knowledge* (Berkeley: University of California Press, 1997), Introduction; J. M. Hodge, G. Hodl and M. Kopf, *Developing Africa: Concepts and Practices in Twentieth Century Colonialism* (Manchester: Manchester University Press, 2014), Introduction; J. M. Hodge, *Triumph of the Expert: Agrarian Doctrines of Development and the Legacies of British Colonialism* (Ohio: Ohio University Press, 2007), Conclusion.
7. A. Staples, *The Birth of Development: How the World Bank, Food and Agriculture Organization, and World Health Organization Changed the World, 1945–1965* (Ohio: Kent State University Press, 2006); G. Rist, *The History of Development: From Western Origins to Global Faith* (New York: Zed Books, 1999); Kiely, *Politics of Labour*.
8. M. Worboys, "Science and British Colonial Imperialism, 1895–1940" (DPhil, University of Sussex, 1979); Hodge, *Triumph of the Expert*; R. Drayton, *Nature's Government: Science, Imperial Britain and the "Improvement" of the World* (London: Yale University Press, 2000).
9. Constantine, *British Colonial Development Policy*; Havinden and Meredith, *Colonialism and Development*, ch. 7.
10. D. Neill, *Networks in Tropical Medicine: Internationalism, Colonialism and the Rise of a Medical Specialty, 1890–1930* (Stanford: Stanford University Press, 2012); K. A. Hoppe, *Lords of the Fly: Sleeping Sickness Control in British East Africa, 1900–1960* (Westport: Praeger, 2003).
11. D. Anderson, "Depression, dust bowl, demography and drought: the colonial state and soil conservation in East Africa during the 1930s", *African Affairs* 83 (1984), 321–343; W. Beinart and J. McGregor (eds), *Social History and African Environments* (Oxford: James Currey, 2003); K. Showers, "Soil erosion in the Kingdom of Lesotho: origins and colonial response, 1830s–1950s", *Journal of Southern African Studies* 15 (1989), 263–286.

INTRODUCTION

12 The history of medicine is an exception, M. Jones, "A 'Textbook Pattern'? Malaria control and eradication in Jamaica, 1910–1965", *Medical History* 57(3) (2013), 397–419.
13 Staples, *Birth of Development*; Rist, *History of Development*.
14 Scott, *Seeing Like a State*; J. Ferguson, *The Anti-Politics Machine: "Development", "Depoliticization" and Bureaucratic Power in Lesotho* (Cambridge: Cambridge University Press, 1990); Cooper and Packard, *International Development and the Social Sciences*.
15 C. Bonneuil, "Development as experiment: science and state-building in late colonial and post-colonial Africa, 1930–1970", *Osiris* 15 (2000), 1501–1520; V. Bernal, "Colonial moral economy and the discipline of development|: the Gezira scheme and 'Modern' Sudan", *Cultural Anthropology* 12(4) (1997), 447–479; M. W. Ertsen, *Improvising Planned Development on the Gezira Plain, 1900–1980* (New York: Palgrave Macmillan, 2016); J. Koponen queries the claim that the Groundnut Scheme is typical of post-war development, "From dead end to new lease of life: development in South-Eastern Tanganyika from the late 1930s to the 1950s", in Hodge, Hodl and Kopf, *Developing Africa*.
16 Hodge, *Triumph of the Expert*.
17 M. M. van Beusekom, *Negotiating Development: African Farmers and Colonial Experts at the Office du Niger, 1920–1960* (Westport: Heinemann, 2002).
18 J. McCracken, "Experts and expertise in Colonial Malawi", *African Affairs* 81 (1982), 101–116.
19 D. Mehos and S. Moon, "The uses of portability: circulating experts in the technopolitics of Cold War and decolonization", in G. Hecht (ed.), *Entangled Geographies: Empire and Technopolitics in the Global Cold War* (Cambridge, Mass.: MIT Press, 2011); H. J. Hoag and M. B. Ohman, "Turning water into power: debates over the development of Tanzania's Rufiji River Basin, 1945–1985", *Technology and Culture* 49 (2008), 624–651.
20 G. Carswell, "Soil conservation policies in colonial Kigezi, Uganda: Successful implementation and an absence of resistance", in Beinart and McGregor, *Social History and African Environments*.
21 Hodge, *Triumph of the Expert*, ch. 5; H. Tilley, "African environments & environmental sciences: the African Research Survey, ecological paradigms & British Colonial development, 1920–1940", in Beinart and McGregor, *Social History and African Environments*; P. Anker, *Imperial Ecology: Environmental Order in the British Empire, 1895–1945* (Harvard: Harvard University Press, 2001).
22 Mehos and Moon, "The uses of portability".
23 L. A. Reynolds and E. M. Tansey (eds), "British contributions to medical research and education in Africa after the Second World War", *Wellcome Witnesses to the Twentieth Century* 10 (2001), pp. 29 and 61.
24 T. Wilkie, *British Science and Politics Since 1945* (Oxford: Basil Blackwell, 1991), p. 11; T. DeJager, "Pure science and practical interests: the origins of the Agricultural Research Council, 1930–1937", *Minerva* 31 (1993), 129–140; A. Hull, "War of words: the public science of the British scientific community and the origins of the Department of Scientific and Industrial Research, 1914–16", *British Journal for the History of Science* 32(4) (1999), 461–481.
25 Constantine, *British Colonial Development Policy*, ch. 7; Morgan, *Official History of Colonial Development*, vol. 1, ch. 4; Goldsworthy, *Colonial Issues*, p. 11; Havinden and Meredith, *Colonialism and Development*, pp. 199–205; Lee and Petter, *The Colonial Office, War and Development Policy*; Ashton and Stockwell, *Imperial Policy*; R. Hyam, *The Labour Government and the End of Empire, 1945–1951* (London: HMSO, 1992); Wm. Roger Louis, *Imperialism at Bay, 1941–1945: The United States and the Decolonisation of the British Empire* (Oxford: Oxford University Press, 1977), pp. 101–103.
26 The remaining British colonies were Malaya, Singapore, Brunei, North Borneo, Sarawak, Hong Kong, the Maldive Islands, Mauritius, the Aden Protectorate,

Cyprus, Malta, Bermuda, Gibraltar, the Seychelles, Fiji, Tonga, the Gilbert Islands, the Soloman Islands, New Hebrides, Ellice Island, St Helena, Ascension Island, Tristan da Cunha, South Georgia, South Sandwich Island, the Falkland Islands and Pitcairn Island.

27 Constantine, *British Colonial Development Policy*, pp. 258–259.
28 L. J. Butler, *Industrialisation and the British Colonial State: West Africa, 1939–1951* (London: Frank Cass, 1997).
29 Kiely, *Politics of Labour*, pp. 5–6; datt Tewarie and Hosein, *Trade Investment*, pp. 2–3; Payne and Sutton, *Charting Caribbean Development*, pp. 2–3; R. Bernal, "The Great Depression, colonial policy and industrialization in Jamaica", *Social and Economic Studies* 37(1/2) (1988), 33–64; see also course readers such as D. Pantin (ed.), *The Caribbean Economy: A Reader* (Kingston: Ian Randle, 2005).
30 Some of the most comprehensive accounts on the 1930s include N. Bolland, *On the March: Labour Rebellions in the British Caribbean, 1934–39* (Kingston: Ian Randle, 1995) and N. Bolland, *The Politics of Labour in the British Caribbean: The Social Origins of Authoritarianism and Democracy in the Labour Movement* (Kingston: Ian Randle, 2001).
31 J. Parker, *Brother's Keeper: The United States, Race and Empire in the British Caribbean, 1937–1962* (Oxford: Oxford University Press, 2008), p. 41.

CHAPTER ONE

New uses for sugar

On the 22 June 1937, Royal Marines from HMS *Ajax* landed at Pointe-à-Pierre in the south of Trinidad. The navy was responding to a request from the Governor for help to suppress riots that had resulted in the deaths of twelve people. *The Times* reported, 'One hundred and fifty marines and blue jackets from HMS *Ajax* are setting up machine-guns to protect the oil fields.'[1] Another navy ship, HMS *Exeter*, arrived at Trinidad the following day.[2] Whilst the violent protests that gripped the island had subsided by the 6 July, three weeks later, a crowd attacked Government House in Bridgetown, Barbados. Four days of unrest followed across the sugar estates of the island, including attacks on shops and lorries and instances of arson, and the Royal Navy were called again. The next year, police fired on a group of protestors at a sugar estate in Frome, Jamaica, leading to a period of violence in the colony. This time the British government responded by appointing a Royal Commission, headed by Lord Moyne, to investigate the conditions that had provoked Caribbean populations to protest on such a scale.

The riots that occurred in the British West Indian colonies during the 1930s have been endowed with much significance by both historians of British imperialism and historians of the Caribbean. Accounts of imperial policy tell how these events were crucial in allowing the Secretary of State for the Colonies, Malcolm MacDonald, to get his way in passing the Colonial Development and Welfare Act of 1940.[3] This Act is considered a turning point in colonial policy as it marked a shift to a more assertive, interventionist form of imperialism that aimed to transform Britain's colonies through development. For historians of the Caribbean, the strikes and riots of the interwar period are a defining moment on the journey towards political independence.[4]

These widespread instances of rebellion illustrate the agency of the subject populations of the British West Indies as people seized the opportunity to protest their grievances over issues such as the slow pace of political change, low wages, inadequate food and housing and the racism they experienced from their employers.[5]

The consequences for the British Caribbean of new legislation for development have been largely unexplored.[6] Almost without exception, we are told only that policy after 1940 for the British West Indies was dictated by the report of the Moyne Commission. In fact, the Colonial Office in London conceived a radical plan for the economic development of the British West Indies that marked a major departure both from previous approaches and from the recommendations of the recent Royal Commission. This policy sought a new and permanent solution to the problem of the low price for sugar that officials considered to be at the root of much unrest. For officials, the lesson of the Great Depression was that profits in the sugar industry could not be maintained on the basis of continuing increases in the volume of production. A new era of prosperity was possible, however, if cane sugar could be reinvented as a raw material for the expanding field of synthetic manufacturing. As chemical companies developed new plastics and medical products, there was increasing demand for supplies of cheap and plentiful starting materials. The Colonial Office decided a programme of scientific research was needed to transform sugar from foodstuff to industrial starting compound. Laboratory investigation was endowed with the power to reverse the long decline of the Caribbean.

This chapter will show how concerns at the Colonial Office around 1940 about the economic future of the British West Indies were expressed as concerns about the future of the sugar industry. While distress was not limited to workers in this industry, and sugar was no longer the principal export of all British Caribbean colonies, it was conditions in this industry that frequently drew the greatest criticism. In addition, the sugar industry was still the biggest employer in the British colonies of the region and when discontent spread amongst workers on the estates it threatened the stability of entire territories. British officials sought a way to revive the fortunes of the Caribbean sugar industry so as to placate colonial agitators and critical foreign governments in the short term and return economic prosperity to this region of the Colonial Empire in the longer term.

An industry in decline

Questions were raised about the long-term future of the Caribbean sugar industry from at least the 1890s, and then in 1934 the price of sugar dropped to an unprecedented low. Officials at the Colonial Office perceived the crisis of the interwar years as different from previous episodes of price instability, believing the world sugar market had now reached the point of saturation. Since profits and wages could no longer be maintained through increases in production, a bleak future existed for the Caribbean. Both the character of the crisis and the timing of it led to a break with previous policy and a search for a new and different solution to the problem of Caribbean sugar.

In the eighteenth century, the West Indian colonies were said to be the richest part of the British Empire, and in 1770 it was estimated that the annual profits from Caribbean sugar were £1.7 million.[7] Sugar from British imperial sources was privileged in the British market from the beginning. From 1651, the Navigation Acts restricted foreign imports to England and its colonies by dictating that only English ships could take goods to the ports of these places. Since the Navigation Acts prevented the movement of English goods directly to foreign ports, they were initially unpopular with sugar planters who wanted access to lucrative foreign markets. Planters were compensated for this loss of trade with other nations through the near-monopoly of the English market.[8] Preferential tariffs were introduced from 1651, with the duties on foreign sugar rising from 270 per cent to 340 per cent of that on West Indies sugar by 1705.[9] In the eighteenth century, episodes during which planters experienced falling profits and production were followed by periods of recovery, and the overall trend was of rising sugar consumption in Britain. Permanent difficulties in the sugar industry of the British West Indies did not become apparent until after 1815.[10]

Decline in the value of sugar from the British West Indies began to occur after emancipation, but was the result of a complex of factors rather than the end of slavery alone. Advocates of free trade first brought an end to preferential tariffs for empire sugar, before duty on sugar was removed altogether in 1874[11] and the Navigation Acts suspended in 1849.[12] As the price for sugar fell, consumption increased sharply, however, from 18 lb per head in Britain in 1800–1809 to 84.7 lb by 1900–1909.[13] Overall, between the 1840s and 1860s West Indian manufacturers saw a decline in value of around 6 per cent but production increases of around 45 per cent.[14] Historians have shown that the post-emancipation pattern of production varied considerably between the colonies of the British West Indies. In Barbados, production grew substantially, with the same trend occurring to a lesser extent in Guiana, Trinidad and St Kitts. Sugar planters in Guiana, Jamaica and Trinidad secured a new source of cheap labour from East Asia with the introduction of indentureship in the 1840s, and wages for sugar labourers generally were kept low through the limited availability of land for peasant agriculture, leaving many people with little alternative but to work on the estates.[15] The maintenance of a source of cheap labour did not prevent rapid decline in the sugar industry in Jamaica between 1840 and 1860, however, although the consequences for the island's economy were moderated somewhat by the production of large amounts of rum.[16]

The factor that caused the greatest problems in the nineteenth century was increasing competition from beet sugar grown in Europe and, to a lesser extent, the growing market share of cane sugar from Java and Cuba. From 1850 to 1900, beet expanded its share of the world market from 16 per cent to 65 per cent, stimulated by the provision of bounties on beet sugar exports which were particularly generous in the case of Germany and Austria, and the introduction of improved, high-yielding varieties of beet.[17] By the 1890s, the British West Indies sugar industry was considered to be in the midst of severe crisis. A slump in price to 10 shillings per cwt (from a high of 97 shillings per cwt in 1814)[18] led to the abandonment of estates, low wages and riots in St Kitts and Guiana in 1896.[19] In that year a West India Royal Commission was appointed to consider the claim that the sugar industry of the British West Indies could only survive in the future with assistance from Britain.[20] In its report, the commission, led by Sir Henry Norman, made a number of recommendations that encompassed the need to have greater diversity of economic activity in Britain's Caribbean territories, including the promotion of peasant agriculture.

Concern about the decline of the West Indies had wider ramifications at the end of the nineteenth century, leading some politicians to advocate a shift in imperial approach towards 'constructive imperialism' in which tariff reform would improve empire trade and provide revenue for initiatives at home. The Secretary of State for the Colonies Joseph Chamberlain is the most famous advocate of a departure from the strict principles of free trade in favour of the promotion of imperial interests. Chamberlain hoped to counter the bounty system used by European countries producing beet sugar by introducing duties on foreign sugar imports. He also asserted that the development of the full economic potential of the colonies would require loans and grants from Britain as a stimulus to greater private investment.[21]

The full ambitions of Chamberlain and the Norman Commission for financial assistance for the West Indies were not realised. The funds that were raised included £250,000 to establish an agricultural bank for sugar planters, a road-building grant for Dominica, and money for establishing peasant smallholdings in St Vincent.[22] The most notable outcome of the recommendations of the West India report was the creation of the Imperial Department of Agriculture in Barbados in 1898, headed by Daniel Morris, previously the assistant director of Kew Gardens in London.[23] A programme of sugar cane breeding aimed at developing higher-yielding varieties of cane was initiated at the department, for which Morris recruited a recent Cambridge graduate, Frank Stockdale.[24] Disease-resistant, high-yielding varieties of cane such as B111 (where the 'B' designated a variety developed in Barbados) were developed through this programme.[25] Apart from this, Britain finally persuaded European countries to abolish subsidies for beet sugar through the Brussels agreement of 1902, although the return to higher prices for cane sugar producers resulted more from a general upturn in commodity prices.[26] Jamaica and Trinidad also saw increases in foreign trade between 1890 and 1914 because of expanding production and export of products other than sugar: bananas in the case of Jamaica and cocoa in the case of Trinidad.[27]

Overproduction in the cane sugar industry increasingly became a problem after the restoration of imperial preference in 1919. The Caribbean colonies experienced substantial increases in production made possible by the planting of improved cane sugar varieties and more efficient methods of extraction in the sugar factories.[28] The generation of a significant surplus on the world market and competition from beet sugar meant that the open market price in London of cane sugar dropped rapidly from 25s 9d per cwt in 1923 to only 8s 3d by December 1929. The British government responded by appointing another commission of enquiry headed by the Fabian socialist Sydney Olivier. Olivier travelled to the British Caribbean colonies in 1929 and 1930 accompanied by the economist and Colonial Office administrator Sydney Caine.[29]

Olivier's 1929 report warned that the cane sugar industry was entirely dependent for its survival on the preference given to empire sugar by Britain, and if this was removed the result would be social disaster in the British West Indies.[30] The imperial preference of around £3 15s per ton that was introduced after the First World War was supplemented after the Olivier Report by a system of colonial sugar certificates with a value of around £1 per ton. The result was that the price of £11 5s received by colonial producers now comprised 40 per cent of assistance. Preferences had the effect of increasing exports of West Indian sugar to the British market at the expense of foreign sugar during the 1930s.[31] Sharply falling prices for commodities during the Depression exacerbated the problem of low price, however, and by 1934, raw sugar had fallen to 3s 10½d per cwt.[32] In an attempt to compensate, sugar manufacturers continued to increase production, which in some islands doubled.[33] Manufacturers in Trinidad introduced new machinery in the field and artificial manure, and deployed improved cane varieties over a greater area.[34] Smaller estates underwent consolidation by firms such as Booker Bros in British Guiana, and Tate & Lyle in Trinidad and Jamaica. Between 1900 and 1950 the number of sugar factories in operation in Trinidad fell from fifty to eight.[35]

In an effort to curb overproduction and an attendant fall in price, the International Sugar Agreement was brokered in 1937 that set quotas for sugar imports. The international character of the agreement was praised at the time for its role in acting as some check against the trend of economic nationalism and protectionism during the Depression.[36] Firms complained, however, about the level at which quotas had been set.[37] Preferences for empire sugar were criticised for the expense to the Treasury and the resultant greater price for the consumer. Some believed that the cane sugar industry should be left to undergo a natural contraction, with all the devastating implications for Caribbean populations, as expressed by *The Economist* in 1930, 'All attempts to artificially impede the restoration of lost equilibrium between these two factors [supply and demand] are useless in the long run, and when they take the form of State assistance, they are doubly objectionable, for they further distort the situation.'[38] While government support for the sugar industry had its critics, officials at the Colonial Office would not contemplate withdrawing assistance for West Indian sugar producers because of the social and political ramifications of a failure of the industry. Britain received a clear warning of the potential consequences of continuing and unchecked decline of the sugar industry during the latter part of the 1930s.

Hunger marches and riots

The predictions made by Olivier in 1929 of a crisis for the Caribbean became reality during the course of the 1930s. A fall in the price of sugar was followed by strikes and protests amongst sugar workers in Trinidad and British Guiana in 1934, in St Kitts, St Vincent and St Lucia in 1935 and then island-wide violence on an unprecedented scale in Trinidad and Barbados in 1937 and Jamaica in 1938. There were official enquiries into the disturbances that occurred in Trinidad, Barbados and Jamaica, and in 1938 a Royal Commission was appointed to investigate the problems of the British West Indies as a whole. These official reports, along with newspaper articles and books such as W. M. Macmillan's *Warning from the West Indies*, brought the plight of the populations of the British West Indies colonies to wider attention. Revelations about the extent of deprivation were fuel for critics of British imperialism, especially those in the US, and this fact enabled the Colonial Office to persuade the Treasury of the necessity of a new Colonial Development and Welfare (CDW) Act in 1940.

The issue that lay at the heart of the riots that had occurred across the Caribbean was a matter of debate. A number of political leaders had emerged in the British West Indian colonies after the First World War and the ability of these individuals to inspire their supporters to violent protest caused alarm amongst colonial governments. British business leaders with interests in the region, such as sugar manufacturers and oil producers, claimed the influence of communists was responsible for strikes.[39] After seeing the conclusions of the various investigations into labour unrest and speaking with colonial governors, the Colonial Office concluded that the main factor that prompted workers to protest across so many of the territories of the Caribbean was the impossibility of living on the pay received. Much of the work on the sugar estates was seasonal and people were paid by task for weeding and planting in the fields, or if cutting cane during the harvest, by weight. In Trinidad, 56 per cent of cane was grown on estates and 44 per cent by independent farmers who then sold their crop to sugar factories.[40] The wages or rates paid for this work were increasingly inadequate in the face of rising food prices during the 1930s, and the difficulty in making ends meet was exacerbated by a lack of full-time employment. The high cost of imported food supplies and the low level of self-sufficiency in growing provisions in the colonies of the British Caribbean meant that the issue that seems to have moved such large numbers of people to violent protest across the region was hunger. The Governor of Trinidad, Sir Murchison Fletcher, wrote after the disturbances on that island to say,

> the immediate origin of the trouble is undoubtedly to be found in economic pressure. Wages of the lowest paid labour have at the best of times given little more that bare subsistence, and for some months past prices have been rising steadily. It is estimated that the increase of the cost of living above the normal level is now in the neighbourhood of 17%.[41]

Official investigators stated that effective labour organisation would have allowed workers to negotiate with their employers on the issue of wages and hence avoid recourse to violence. Labourers reported their fear, however, that complaint would result in dismissal since high levels of unemployment meant a pool of workers ready to take the place of seemingly troublesome individuals. Overpopulation in Jamaica and Barbados worsened during the 1930s as migrants returned from work on the Panama Canal or from the US and Cuba on the loss of their jobs as the Great Depression deepened.[42] One sugar worker who wrote to express his grievances to the Governor of Barbados told how it was pointless for labourers to ask for higher wages on the estate he worked on since 'Barbados is so thickly populated and the money man knows that if one refuses he can get fifty to take his place as they are starving.'[43] Those who complained about the unfairness of pay arrangements noted that whilst field labourers struggled to feed their families, profits were still being generated by the sugar manufacturers and bonuses were being paid to managers. 'Grevious citizen' in Barbados wrote to the Governor, 'Mr Taylor at Wakefield has a dog, it gets beef steak three times a week, Ovaltine and other things and we the working man don't get even good salt fish.'[44]

The disturbances in Trinidad and Barbados in 1937 did not begin amongst sugar workers, but these workers were moved to riot in large numbers as word of unrest spread. In Trinidad, trouble on the island began in the Forest Reserves oilfield belonging to Trinidad Leaseholds, Trinidad's largest oil producer. On the evening of 19 June, the police attended a meeting of around 200–300 oilfield workers with the intention of arresting the leader of a recent strike, Uriah Butler. Butler was a highly charismatic orator who had formed a left-wing political party in 1936, the British Empire Workers and Citizens Home Rule Party, which had a committed following in the oilfields in the south of Trinidad.[45] In the commotion that ensued after police attempted to break up the meeting of strikers, Corporal Charles King of the Fyzabad police was pursued by a group of women, either fell or was pushed from a window, and was set alight. Within two days, strikes and protests had swept the island, spreading from the oil fields of the south to the sugar and cocoa estates in central Trinidad and then to workers in the capital, Port-of-Spain. Three weeks of island-wide violent protests followed, and by 6 July, fourteen people were dead, fifty-nine wounded and hundreds of people had been arrested.

Around three weeks later, on the evening of 26 July, violence broke out in Bridgetown, Barbados, when a crowd descended on Government House to protest about the threat of deportation levelled at Clement Payne, the founder member of the Barbados Labour Union. The crowd damaged street lights and parked cars and threw stones at the police.[46] Payne had arrived in Barbados from Trinidad in March 1937, where he had been involved in the work of the Negro Welfare Cultural and Social Association. The NWCSA, formed in 1935, was dedicated to protesting the grievances of Trinidadians and campaigning against events that occurred on the international stage, such as the Italian invasion of Ethiopia. Whilst in Barbados, Payne had addressed political meetings, urging Barbadian workers to rise up against their employers, and he had come to the attention of the Barbados police. The order to deport Payne was officially made on the basis that he had falsely declared himself to be Barbadian on entry to the colony, when in fact he had been born in Trinidad, but seems more likely to have been the result of a sense that he was a trouble-maker.

While the attempt to remove Payne led in the first instance to an outbreak of violence in the capital, by 28 July there were reports that sugar workers in the rural parishes had been called out to strike, and in the afternoon a group of around 200 people descended on the Brighton and Carmichael plantations, raiding fields for food and stopping people from working. Strikes spread across the whole of the island, shops and lorries were attacked and there were further instances of food theft and arson. By the end of the disturbances, four days later, fourteen people were dead and forty-seven injured. The official enquiry into the unrest noted 'The lawless acts committed in the country were more purposive than those committed in Bridgetown, and it would appear that hunger or the fear of hunger, coupled with the news of the disturbances in Bridgetown were the chief causes of the outbreaks in the country districts.'[47] In the case of Barbados and Trinidad, armed police and the Volunteer Force initially dealt with the disturbances before the governors of the two islands requested help from the Royal Navy, and sailors and marines landed from HMS *Ajax* and HMS *Exeter* in the case of Trinidad and HMS *Apollo* in the case of Barbados.

These two episodes of unrest were set off by clashes between police and the supporters of Butler in Trinidad and Payne in Barbados. Both men were political dissenters who mobilised people to join forces and protest against the abuses they endured. The broader context to the actions of individuals such as Butler and Payne and their followers was one of increasing awareness and anger about racial injustice, fuelled by the experiences of black soldiers such as Butler who fought for the British during the First World War, the rise of the pan-African movement and fury over the perceived betrayal of Ethiopia when Britain recognised Italian control of the country in 1938. Increasing political involvement was also found amongst the East Indian population of Trinidad, as exemplified by the radical barrister Adrian Cola Rienzi, who set up the Trinidad Citizens League in 1935 from which Butler left to form his own British Empire Workers and Citizens Home Rule Party. Whilst the interwar period saw increasing levels of political participation and organisation amongst the populations of the Caribbean, the rapid spread of violent protest across so many workers in the British colonies in 1937 and 1938 was the expression of widespread anger about economic hardship. Colonial Office officials in London believed that the imprisonment of communist sympathisers, or the creation of a permanent military station as oil company managers had requested for Trinidad, were unlikely to resolve the problem. The conclusion of the official enquiries into unrest in Trinidad, Barbados and Jamaica was the same: the threat of further violence would not be removed without measures to alleviate the deprivation that existed in every colony. According to the commission of enquiry appointed to investigate the cause of the unrest in Barbados,

> we must guard against conveying the impression that we think that the disturbances were a mere flash in the pan, a spark of revolt which might have been extinguished by such measures [as a stronger police response]. On the contrary it is our considered opinion after surveying the whole field that there was a large accumulation of explosive matter in the island to which the Payne incident only served as detonator. That the real cause of the disturbances was in fact economic can we think be convincingly shown; further we are of opinion that the conditions which rendered this culmination possible still exist and demand immediate treatment.[48]

The most visible and obvious sign of the deprivation endured by workers in the Caribbean was the squalid housing conditions provided for them, which were condemned by every official investigator dispatched to the region. The commission of enquiry into the disturbances in Trinidad and Tobago generally avoided strong, direct criticism of the treatment of labour by the oil companies, but in the case of the sugar industry pointed to a clear connection between the poor-quality housing on the estates and the discontent of sugar workers.[49] In the autumn of 1937, the Labour MP and union official John Jagger travelled from Manchester to Trinidad to attend an arbitration tribunal, convened to determine whether there should be an increase in wages in the oil industry. Jagger kept a diary of his time on the island between November 1937 and February 1938 in which he described the conditions he saw there. On visiting the barracks that housed sugar workers on the estates of central Trinidad, Jagger wrote,

> I thought that I had seen the worst that anyone could show in housing conditions when we went round the oil fields, but I must hand it to the sugar firms for absolute filth, ignoring of any kind of sanitary conditions whatever, and for general misery. Words fail me when I try to describe the conditions we saw – open drains with green slime a couple of inches thick flowing both behind and before the wretched wooden shacks in which the workers were living. Three single men to a room 10ft square, and a room of white washed boards plus a galvanised roof which had rusted till the sunlight and the rain could enter at almost every point. Similar rooms where families of five or ten persons of, pigged in, regardless of age or sex. Filthy latrines where chloride of lime only served to add one more obnoxious smell to the rest, and excrement flowing down the open drains. Perhaps most dreadful of all were the round iron cisterns at each door containing the drinking water for the residents, in which

were every creeping and crawling thing imaginable, plus endless masses of mosquito larvae.[50]

The scene sat in strong contrast to the manicured golf course and beautiful bungalows for the European staff of the sugar estates.[51]

In May 1938 violence erupted in Jamaica on a sugar estate at Frome recently acquired by the sugar company Tate & Lyle. Some 3,000 workers converged on the estate to agitate for better wages, and after employees in the main office of the company were attacked, police fired their rifles into the crowd and killed four protestors.[52] This episode received a great deal of coverage in the British press and an article in The Times reported that, alongside low wages, the issue that had led to the riot was the refusal of Jamaican workers to continue to accept the appalling housing conditions provided on the sugar estates.[53] In a letter to the paper, Leonard Lyle, President of Tate & Lyle, rejected the argument that conditions on the company's Caribbean sugar estates were in need of improvement on the grounds that it was wrong to consider Jamaican workers as requiring the same standard of working and living conditions as British labour: 'We must be careful to remember that the West Indian labourer does not even remotely resemble the English labourer, either in his mode of life or his mentality.'[54] Lyle denied the claim that riots in Jamaica stemmed from lack of employment and poor wages and housing, and he continued by stating that 'by no means a small proportion of our British troubles overseas are caused by the sinister influences of the communists'.[55] The Colonial Office, however, was prompted by the Jamaican unrest to appoint a Royal Commission for a comprehensive investigation of social and economic conditions across the British Caribbean and started to prepare new legislation that would provide the colonies with increased development grants, including money for welfare reform.

The reform of policy

Concern had existed about the decline of the British West Indies for a considerable period of time, but events during the 1930s had greater impact than previous episodes of depression and unrest. The disturbances in the British Caribbean during the 1930s, along with those in Northern Rhodesia and Mauritius, occurred at a critical juncture for Britain.[56] The integrity of the British Empire was under threat from Germany's demands for the return of its former colonies and Japanese expansion in south-east Asia.[57] In addition, there were vocal critics of imperialism in the US.[58] With the outbreak of war, the Colonial Office argued that it was imperative that Britain take action to secure the continuing loyalty of colonial peoples and ensure there was no interruption to colonial production. Trinidad was of particular strategic importance to Britain as it was the largest source of aviation fuel in the empire; a high-octane plant had been built by Trinidad Leaseholds at their refinery at Pointe-à-Pierre in the south of the island in 1937.[59] Beyond the need to prevent colonial revolt, officials feared that any accusation that Britain was incapable of ensuring reasonable levels of social provision for its colonial subjects raised the possibility that after the war Britain's colonies might be removed altogether and placed under League of Nations mandate.[60]

The importance of the unrest in the West Indies during the 1930s for the reform of legislation that led to the 1940 CDW Act has received substantial scholarly attention.[61] The aim here is to show how the new CDW Act was used by the Colonial Office as the opportunity, and the means, to find a long-term answer to the problems of the West Indian sugar industry. In the aftermath of the Caribbean riots, oversupply of sugar on the world market was identified as a major issue.[62] The vision of economic development that the Colonial Office subsequently produced for the Caribbean was one in which sugar was to be transformed from a low-value foodstuff to a high-value industrial raw material. Frank Stockdale, appointed Comptroller for West Indian Welfare and Development in 1940, linked the need to deal with a future sugar surplus to the alleviation of social problems in the British Caribbean:[63]

> We think, at present, of sugar only as a food but the field of consumption will have to be extended, if the post war situation is not to find us with no alternative but restriction, increased unemployment, distress and misery in the West Indian colonies.[64]

The solution that was devised by officials in the Economics Department made use of new provision for scientific research created with the passing of the CDW Act of 1940.

The announcement that Britain was making a substantial new commitment to economic and social development and scientific research through the 1940 CDW Act was timed to coincide with the publication of the recommendations of the West India Royal Commission, the findings of which were expected to cause great embarrassment to the British government. The Colonial Office appointed the West India Royal Commission, headed by Walter Edward Guinness (Lord Moyne) in the wake of the riots in Jamaica. The dispatch of the Commission to the Caribbean had the short-term function of providing some evidence that Britain was taking measures to investigate conditions in its colonies, in an attempt to counter domestic and foreign criticism.[65]

The Commission included experts on economics, education, social reform, trade unions and agriculture, reflecting its remit to undertake a comprehensive investigation of the material, political and social issues affecting the British Caribbean.[66] Its role as a public gesture that Britain was not indifferent to the grievances of its colonial subjects was executed by giving a voice to groups and individuals in the British West Indies themselves. During its tour of the colonies of the British West Indies and at its London meetings the Commission received 789 memoranda and 300 other communications and saw 370 witnesses or groups of witnesses.[67] When the final report was presented in 1940, it provided a detailed and at points candid description of the hardships endured by the populations of the British West Indies.

While waiting for the submission of the report, the Colonial Office set about drafting reforms to its 1929 Colonial Development Act. The priority of that legislation had been the encouragement of projects of economic development in the colonies as a way to alleviate domestic unemployment. The fact that the 1929 Act had failed to stimulate widespread colonial development, and indeed many colonial economies were now in worse shape than they had been before the Great Depression, prompted a review of policy in June 1938, and by February 1940 a draft of a new CDW Act had been prepared.[68] The reformed CDW Act included an expanded fund for development of £5 million pa (an increase from the annual £1 million allocation given in the 1929 Act), free grants rather than loans to the colonies, including grants for projects of social improvement, and the creation of an annual fund of £500,000 for scientific research.

The challenge facing Malcolm MacDonald was to persuade the Treasury to endorse the CDW Act when Britain had recently entered a war. In his communications with the Chancellor of the Exchequer John Simon, MacDonald raised the spectre of further episodes of colonial rebellion, with a clear allusion to recent events in Jamaica. The proposed Act was essential, according to MacDonald, in sending a message to aggrieved populations that the problems they faced were going to be addressed. Without such measures, there was a real danger that workers in the colonies, particularly those who possessed a wider understanding of the injustice of their situation, would cause trouble again.

> We know what form such trouble takes. On some pretext or other there is a strike accompanied by rioting and sometimes even by murder; as often as not our police have to fire on the crowds; troops are even called out; and occasionally it is necessary to summon a war-ship in aid to land marines.[69]

This would be a disaster for Britain during the war. Large-scale colonial unrest could interrupt the production of essential materials, the Royal Navy could not spare ships to contain riots, and violent protest provided propaganda for Britain's enemies. MacDonald warned that the shooting of protestors as had occurred at Frome in Jamaica in 1938 would be used by Germany as evidence of Britain's incompetence and willingness to act oppressively towards its colonial subjects. This could have damaging consequences for the relationship between Britain and the US. MacDonald warned the Treasury not to think that the peace that currently prevailed in the colonies was anything but a temporary lull, produced in the case of the West Indies by the dispatch of the Moyne Commission to the region.[70]

It has been observed that Royal Commissions can function as a way to bring episodes of public controversy to an end and that while commissions are presented as evidence of government taking action, they can actually work to quietly shelve issues.[71] In the case of the Moyne Commission the Colonial Office was clear that the launch of this independent investigation was not going to bring resolution to the angry debate about Britain's responsibilities towards its colonial possessions. It was anticipated that the final report of the West Indies Royal Commission would generate a great deal of criticism of British imperial rule, and MacDonald told the Treasury this needed to be offset by the quick announcement of new plans and money to address the distress that would no doubt be detailed by the Commission. The Treasury was persuaded and the CDW Act was passed in February 1940. The announcement of Britain's new provision for colonial development was made on 16 February to coincide with the publication of the recommendations of the Moyne Commission. The full report was considered so potentially damaging to Britain's reputation that it was not published until after the end of the war in 1945.

War meant there was not a great deal of initial spending from the new CDW Fund and officials instead took the opportunity to debate and develop their proposals for the post-war period. The Act included a commitment of £1.4 million specifically for West Indian problems and the Colonial Development and Welfare Organisation (CDW Org) was created to deal with development plans for this region, headed by Sir Frank Stockdale, who had started his own colonial service career as a sugar cane breeder in Barbados. By August 1940, Stockdale was involved in discussions with members of the Economics Department of the Colonial Office about the future of the West Indian sugar industry. With the outbreak of war the International Sugar Agreement had been suspended and the Ministry of Food controlled all sugar purchasing and wholesale and retail prices from September 1939.[72] Whilst oversupply was not a problem for the time being, officials were certain that 'sugar is going to be a post-war problem of some importance and it would be very useful, if it could be done, to get a flying start'.[73] In their deliberations on the issue of the sugar industry the Economics Department began by considering the recommendations of the Moyne Commission.

The Moyne Report painted a gloomy picture of the future of the Caribbean sugar industry, stating that world markets for tropical commodities no longer offered opportunities for expansion. The difference between the crisis of the 1930s and the Depression of the 1890s was that, 'Then the world demand for almost every tropical product was increasing so rapidly as to outstrip, subject to the ordinary ups and downs of trade, the available supply, and to require for its satisfaction the opening-up, one after other, of new productive areas.'[74] In the late 1930s, however, the world market for sugar was said to have reached the point of saturation. Whilst the issue for tropical commodities generally had become one of oversupply and price depression, the situation with sugar was particularly acute as this industry had been reliant for a long period on subsidies, preferences and protection, all of which had been extended and intensified since the First World War.[75]

Despite the pessimistic projection of the future prospects for sugar, the Moyne Commission's recommendations were modest. It was suggested that the system of colonial sugar certificates could be modified so that preferences were paid out not on a fixed quantity of sugar but on one half of the volume of the total exports of any colony. In addition, it was proposed that a minimum price be set so that if the price of sugar were to fall to a critical level it would trigger an increase in the preference paid out. The aim was to ensure greater stability, and some increase, in the price paid for Caribbean sugar. It was an approach that was merely palliative rather than curative, and it was clear that, in the opinion of the Moyne Commission, the sugar industry of the West Indies could not anticipate a new era of prosperity.

The idea that the world market for sugar as a foodstuff had reached saturation point was an issue taken up by the Economics Department of the Colonial Office in August 1940. Sydney Caine, the office's Finance Adviser and someone with experience of seeing the British West Indies first-hand as part of the Sydney Olivier Commission of Enquiry in 1929, agreed that in the future there was no point encouraging increases in the production of agricultural commodities in anticipation of ever-expanding markets. Instead, he suggested, the Colonial Office should consider helping primary producers take advantage of the increasing demand in wealthy countries for non-foodstuffs such as paper, fibres and fuel.[76] In a break with the recommendations of the Moyne Commission, Caine and his colleagues in the Economics Department of the Colonial Office decided to seize the initiative and find a way to assure a profitable future for cane sugar by establishing it in new and more lucrative markets as a raw material for fuels and synthetic products.[77]

New uses for sugar

In the summer of 1940, officials at the Economics Department discussed the possibility of finding new transport and industrial uses for cane sugar as a solution to the problem of future oversupply. The inspiration for using sucrose to make alcohol-based fuels, and as a raw material for the chemical industry, came from the experiences of the interwar period. The high price of petrol and anxiety about Britain's dependency on imported oil had led government and business to investigate the production of power alcohol after the First World War. Power alcohol was alcohol produced from a fermentable organic product such as potatoes, grapes, sugar or grains that was then blended with petrol for use in vehicles. The interwar period was a high point for the development of alcohol and petrol blends for use in cars as a number of nations in Europe and North America sought self-sufficiency in fuels and a solution to the problem of agricultural surpluses during the Great Depression.

The interwar years had also seen British chemical firms searching for cheap and abundant raw materials for the manufacture of their expanding range of products. These products included synthetic plastics such as polythene, developed by Britain's largest chemical company ICI during the 1930s. For companies such as ICI there were three possible options in terms of raw materials for manufacturing organic chemicals – coal, oil or molasses. In Britain, government legislation had given domestic producers of industrial alcohol an inconvenience allowance intended to be compensation for the stoppages that occurred because of inspection of factories by Excise officials.[78] According to Ronald Weir, 'of the possible raw materials for organic chemicals, molasses "was the least likely of them all"', but the allowance made alcohol from molasses the cheapest raw material available in Britain for the chemical industry.[79]

The most important firm in Britain in terms of producing fermented alcohol from molasses for the manufacture of chemicals was the Distillers Company Ltd (DCL). The success of DCL in producing power alcohol, industrial solvents and other products from molasses encouraged the Colonial Office to consider the potential of a market for industrial products derived from sugar in the early 1940s. Another legacy of sustained interest and activity in the field of power alcohols and organic chemicals during the interwar period was growth in expertise in organic chemistry and fermentation processes amongst British scientists, both in university departments and chemical companies. The endorsement by some eminent chemists of the potential of using sugar to make fuels and synthetic goods was crucial in persuading officials at the Colonial Office of the validity of a programme of research in this area.

The first suggestion raised amongst Colonial Office officials was to use sugar to make power alcohol for use as aviation spirit, thereby meeting a wartime need. Treasury funds were likely to be forthcoming for a proposal to examine alternatives to fuels from oil during wartime because, as noted by Gerald Clauson, Head of the Colonial Office Economics Division, 'the Treasury would be favourable to anything which tried to substitute alcohol for oil, for oil to them is dollars'.[80] In August 1940, Stockdale met with the Fuel Research Board (FRB) of the DSIR in order to solicit their opinion of the proposal to convert Caribbean sugar into alcohol for aviation fuel. The FRB was the most expert government body on the matter of power alcohols in Britain and had produced a series of detailed memoranda between 1920 and 1927. These reports had been prompted by widespread demand for an alternative to imported oil following a steep rise in the price of petrol after the First World War. Fear had existed that Britain was dangerously dependent on supply from just two companies, Royal Dutch Shell and Standard Oil, who were accused of wielding excessive power in the international market for oil. As an alternative, the derivation of liquid fuels from coal was considered to have possibilities but it was feared that existing sources of fossil fuels would be exhausted at some point in the near future. In contrast, alcohols from plant materials had the great advantage of being produced from sources that could be indefinitely replenished.[81]

A desire for greater self-sufficiency in fuels was not restricted to Britain, and a number of European governments initiated national power alcohol programmes during the interwar period. In France a law was passed making the use of alcohol blends by motor vehicles compulsory using alcohol originating from the grape harvest. The aim was to protect a domestic alcohol industry that was considered essential in case France was ever at war again.[82] Interest in the use of agricultural products to make fuels was also high in America. The chemist William Jay Hale advocated the production of power alcohols as a method of using the agricultural surpluses that arose during the Great Depression. Hale christened the use of farm products in the chemical industry as chemurgy. In response to the chemurgic movement, the

US Department of Agriculture created four new research laboratories that dealt with local agricultural surpluses, including the Northern Regional Research Laboratory at Peoria, Illinois, which was to go on to play a crucial role in the development of penicillin.[83]

In Britain, the need for government action to allow cheap imports of alcohol, or cheap domestic production, to meet the future needs for fuel by the British motorist was promoted by the Inter-Departmental Committee, appointed by H. M. Petroleum Executive in 1918. This body asserted a need for government intervention to ensure the quick and efficient resolution of the issue of finding a competitively priced fuel for the motoring public:

> We think that the development of the alcohol industry cannot be left entirely to the chances of private enterprise, individual research and the ordinary play of economic forces. No doubt in the long run, after a tedious process of trial and error, alcohol would find its proper place as a power fuel, but only with the maximum of friction, great fluctuations in price and serious waste of time and energy. The situation needs to be watched continuously and measures taken from time to time to ensure a smooth and rapid adjustment of supply to demand.[84]

The FRB was given responsibility for overseeing scientific research into producing power alcohol at reasonable cost.[85] Sir Frederick Nathan, chair of the Inter-Departmental Committee and who had previously worked at the Nobel Explosives Factory and the Royal Gunpowder Factory, was appointed Power Alcohol Investigation Officer in December 1919. In 1920 the inconvenience allowance for domestic users of alcohol produced from fermentation was extended to power alcohols.

The FRB oversaw a programme of research in the 1920s into two possible routes to alcohol production: chemical transformation and fermentation by microbes. The latter was investigated at a laboratory at the Royal Naval Cordite Factory in Wareham, Dorset, where scientists studied the fermentation of Jerusalem artichokes.[86] A research group had been established at this laboratory during the First World War under Chaim Weizmann, future President of Israel, who was then Reader in Biochemistry at Manchester University. The initial purpose had been to develop the large-scale production of acetone based on a process worked out by Weizmann. Acetone was a key military material used in the production of smokeless cordite, a propellant for shells and bullets favoured by the navy as the absence of smoke meant that it did not allow the position of a ship to be determined.[87] Before the war Britain had imported acetone derived from wood from Austria, Canada and the USA. By 1915, reserve stocks were extremely low and the Admiralty engaged Nathan and William Rintoul of the Nobel Explosives Company to secure new sources of desperately needed chemicals. The search for acetone had not been successful until Weizmann contacted Rintoul offering a new chemical process for its manufacture.[88]

Weizmann's research group included a number of well-known microbiologists who subsequently worked for government or business, including A. C. Thaysen, who was employed by the Colonial Office after the end of the Second World War, and T. K. Walker, who went on to establish the new department of Industrial Fermentation at the College of Technology in Manchester in 1925.[89] According to the fermentation chemist John J. Hastings, who was trained by Walker, 'It was this small band of workers and those they trained after them, who produced a generation of craftsmen in industrial bacteriology, with a skill that made most medical bacteriologists look like plumbers.'[90] Hastings himself went on to work at the Commercial Solvents Corporation, a Liverpool-based manufacturer of industrial alcohol that used Weizmann's process.[91]

Aside from attempting to resolve the technical problems associated with producing alcohol from plant materials, the FRB considered practical and economic issues related to the production and marketing of alcohol. The goal was to find a raw material and method of production that produced alcohol at a cost cheap enough to make a fuel that could compete with petrol. One key problem was obtaining sufficient quantities of a suitable raw material. The FRB dismissed molasses relatively quickly because the producers in the empire were too small and scattered. Jerusalem artichokes and mangolds were considered good potential domestic sources of alcohol, with the additional benefit of also being a source of cellulose for making paper and, in wartime, nitrocellulose.[92] The FRB was concerned, however, that a domestic power alcohol industry could only be sustained by a considerable expansion in the farmland given over to these crops. The FRB estimated that consumption of motor fuel in Britain for 1920 was around 250 million gallons. Fuel based on alcohol would require a crop of mangolds of 25,000,000 tons. In 1919 the total year's crop had been 7,769,000 tons. The increase in farmland to make up the shortfall was estimated by the board to be an additional 810,754 acres.[93]

By 1924, the FRB had concluded that a power alcohol industry based in Britain would find it very difficult to secure substantial volumes of suitable raw materials at a sufficiently low cost to be able to produce a fuel that could compete with petrol, and the FRB made this point to Stockdale in 1940 when he enquired about the potential of using sugar to make alcohol fuels. Despite the reservations of the FRB, however, an example did exist of a commercially successful power alcohol blend that had been marketed in Britain in the 1930s.

Discol was a mix of alcohol and petrol that had been made by DCL using molasses. DCL had overcome the problem of securing a supply of molasses by purchasing a shareholding in the newly formed United Molasses Company, the largest importer of molasses to Britain, and in 1926, research staff at the firm succeeded in producing a form of denatured or absolute alcohol with no water content, thereby overcoming an important technical obstacle to the manufacture of an alcohol for a blended fuel that might prove attractive to the oil companies.[94] This absolute alcohol possessed anti-knock properties and so could compete with new types of petrol containing tetra ethyl lead.[95] From 1933, the Cleveland Petroleum Company marketed a fuel named Cleveland-Discol utilising absolute alcohol manufactured by DCL, and this product proved popular with motorists. The success of Discol prompted Stockdale to suggest that there might be interest amongst Trinidad oil companies in producing a similar alcohol and petrol blend.[96]

DCL proved to be a major source of inspiration to the Economics Department of the Colonial Office in other ways. Stockdale reported to his colleagues that there was increasing demand in Britain for industrial alcohol (methylated spirit) to produce solvents for the manufacture of paints, varnishes and other products. By 1939, DCL was the dominant producer of industrial alcohol in Britain, controlling over four-fifths of the supply. Alcohol derived from molasses formed the basis of an array of new chemical products produced and consumed in Britain and its empire, including plastics such as polyethylene. The largest customers of DCL included ICI, National Benzole and British Celanese, makers of artificial silk and other acetate products.[97] DCL itself had diversified from being a drinks manufacturer, with a focus on whisky production, into a producer of yeasts, motor fuels, ethylene oxide, ethylene glycol, synthetic resins, acetone, aldehydes and ketones, mainly through the strategic acquisition of patents and other businesses.[98] DCL was able to expand its range of products so substantially as alcohol fermented from molasses was a starting material of enormous versatility.[99]

Demand in Britain for raw materials for the chemical industry expanded hugely during the interwar period as the volume and range of chemical products grew. Between 1924 and 1934, the value of the fine chemicals and synthetic organic sector in Britain increased from £2.6 million to £7.7 million.[100] Government had an important role in fostering this growth. As well as providing a discount on industrial alcohol from molasses through the inconvenience allowance, the Safeguarding of Industries Act in 1921 protected domestic manufacturers of synthetic chemicals through tariffs.[101]

Stockdale attempted to persuade his colleagues at the Colonial Office in 1940 that expanding demand for solvents indicated that there was a market for using sugar as a raw material for the chemical industry, and he proposed that the Colonial Office pay for a programme of research.[102] Sugar was presented as a comparable raw material to molasses; it was cheap and plentiful and even had the advantage of greater purity. The suggestion that sugar might be a raw material for fuels and industrial alcohol capitalised upon the unprecedented level of interest in using molasses and vegetable products as raw materials during the interwar period amidst concern about the price of motor fuels and the growth of the chemical industry in Britain. For officials at the Colonial Office it seemed that a fantastic opportunity existed to divert Caribbean sugar to use as an industrial raw material. The fact that DCL's future business as a producer of alcohol from molasses was dependent on continuing government support in the form of the inconvenience allowance did not deter the Colonial Office from deciding to promote sugar cane as a raw material for making fuels and synthetics as a way of resolving the economic problems of the British Caribbean.

The Colonial Products Research Council (CPRC)

Stockdale, Caine and Clauson decided that the next stage in their plans to find new uses for sugar was some scientific investigation, using the Research Fund created as part of the 1940 CDW Act.[103] The idea that the Colonial Office sponsor research into sugar was endorsed, perhaps unsurprisingly, by the scientist they consulted on this point, Sir Norman Haworth, a leading figure in the field of sugar research. Haworth had trained with W. H. Perkin and had held the position of Mason Professor of Chemistry at the University of Birmingham since 1925, where he established a major school of research in carbohydrate chemistry. In 1937 he was awarded the Nobel Prize in chemistry for the development of a process to make synthetic vitamin C.[104]

Haworth was enthusiastic about the suggestion that research could be done with the goal of broadening the use of cane sugar and told the Colonial Office that Philip Lyle from Tate & Lyle had assured him that refined sugar could be made available at a price as low as £7 a ton, making it a raw material for the chemical industry that was competitive with coal and oil. Haworth went on to say that he had recently been in discussion with ICI Dyestuffs Ltd on the issue of using sugar to make synthetic products and a preliminary programme of research had already

Figure 3 Norman Haworth.

been started at the University of Birmingham. Haworth then went on to offer his services to the Colonial Office to undertake further work.[105]

The Colonial Office created a committee of scientists and business representatives to determine other areas where scientific research might be of use, the Scientific Committee for Examining Alternative Uses of Colonial Raw Materials (SCEAUCRM).[106] The committee was told that before the war the problem of oversupply of colonial products had been tackled with international agreements that regulated production but that in the future it was thought more likely that new markets could be found for manufactured goods based on tropical commodities than that new markets would be found for food products in excess supply.[107] The need for research into new uses for colonial foodstuffs was therefore linked to expectations about the longer-term economic development of the Colonies.[108]

One of the first actions of SCEAUCRM was to meet with Chaim Weizmann with regard to a letter from Weizmann to the Secretary of State for the Colonies, Thomas Lloyd.[109] Weizmann had lobbied Thomas Lloyd to follow his instructions for manufacturing high-octane aviation fuel from sugar and molasses.[110] In his communications, Weizmann emphasised the importance of the security that would be afforded to Britain and its empire if self-sufficiency in fuels could be attained. According to Weizmann, the danger facing Britain was that oil would soon dry up in the US and that Britain would become dangerously dependent on supplies from the Middle East.[111] Unlike the contribution made by Haworth, however, Weizmann's intervention did not give the idea of power alcohols greater credibility. Officials dismissed Weizmann on the basis that he had an agenda. John Shuckburgh of the Colonial Office wrote to his colleague Cosmo Parkinson to say, 'I should be sorry to think that Dr Weizmann's motive in raising this question with Lord Lloyd was purely "political" (meaning I suppose, that he wished to gain credit for himself to exploited in Zionist interests after the war).'[112] Weizmann had taken the opportunity in his communications with Lloyd to express his disappointment at Britain's failure to support the formation of the Jewish National Home.[113] During the First World War, Weizmann had gained the trust and respect of Lloyd George for his scientific work and he became an insider amongst politicians and the civil service for a period, gaining support from the British government for his Zionist cause.[114] This time, however, officials declined the offer of assistance from Weizmann and rejected the idea of using sugar and molasses to produce a high-octane aviation fuel.[115]

By September 1941, SCEAUCRM had secured the agreement of the Treasury for the funding of research into a range of tropical products at British universities, and for the formation of a new body, the CPRC, overseen by a salaried director of research.[116] Lord Hankey was offered the chairmanship of the CPRC in November 1941 and the post of director was taken by a chemist from the University of North Wales, Professor John Simonsen.[117] One of the first projects approved by the CPRC was a programme of research into sugar in Haworth's laboratory at Birmingham University.

Conclusion

Histories of the Caribbean colonies from the interwar period to independence from Britain often rely upon a narrative in which the riots in the 1930s prompted an investigation by the Moyne Commission, and the report produced by this body then set out the blueprint for Britain's post-war policy for the region. This chapter has shown that the British response to unrest in the Caribbean in the 1930s was comprised of more than the appointment of the Moyne Commission and the passing of the 1940 CDW Act. The events of the late 1930s prompted a discussion at the Colonial Office about the future of the Caribbean sugar industry that quickly diverged from the recommendations of the Moyne Report for modest adjustments to preference arrangements. The solution to the problems of the sugar industry devised by Clauson, Caine and Stockdale at the Economics Department represented a significant break with the past. The proposals of officials gave a central role to laboratory studies of sugar with the goal of finding new industrial and fuel uses for this commodity. This was a suggestion that emerged from the search for cheap and reliable raw materials as fuels and starting compounds for the expanding British chemical industry that had been under way since the First World War. The Colonial Office ambition was to exploit this need in order to find a permanent solution to the problem of the low price of cane sugar and the social deprivation it produced. The intention was to bring to an end a lengthy period in which the only way government could avert social and political disaster in the British Caribbean was by using public funds to supplement the price paid for sugar.

Whilst officials drew upon the experiences of British firms in the interwar period when promoting the idea that sugar could be a raw material for the chemical industry and to make fuels, the fact that the Colonial Office decided to fund a programme of scientific research into the chemistry of sugar requires further explanation. The CPRC was the product of the Research Fund created as part of the 1940 CDW Act. This substantial research fund marked a wider technocratic turn at the Colonial Office after 1940 in which scientific research was given a privileged role in an invigorated programme of development for Britain's colonies.

Notes

1 *The Times*, "Marines landed at Trinidad: disorders spreading" (23 June 1937), p. 15.
2 *The Times*, "Another warship at Trinidad, three strikers shot" (26 June 1937), p. 13.
3 Constantine, *British Colonial Development Policy*; Havinden and Meredith, *Colonialism and Development*; Morgan, *Official History of Colonial Development*; Butler, *Industrialisation*.
4 Bolland, *On the March*.
5 Bolland, *On the March*.
6 R. Harris, "From miser to spendthrift: public housing and the vulnerability of colonialism in Barbados, 1935–1965", *Journal of Urban History* 33 (2007), 443–466; J. French, "Colonial policy towards women after the 1938 uprising; the case of Jamaica", *Caribbean Quarterly* 34 (1988), 38–61; R. Harris, "Making leeway in the Leewards, 1929–51: the negotiation of colonial development", *The Journal of Imperial and Commonwealth History* 33 (2005), 393–418.
7 J. R. Ward, "The profitability of sugar planting in the British West Indies, 1650–1834", *Economic History Review* 31 (1978), 197–213; K. Morgan, *Slavery, Atlantic Trade and the British Economy, 1660–1800* (Cambridge: Cambridge University Press, 2000), p. 49.
8 R. Davies, "The rise of protection in England, 1689–1786", *The Economic History Review* 19 (1966), 308; R. B. Sheridan, "The Molasses Act and the market strategy of the British sugar planters", *The Journal of Economic History* 17 (1957), 62.

9 D. A. Farnie "The commercial empire of the Atlantic, 1607–1783", *The Economic History Review* 15 (1962), 209.
10 Morgan, *Slavery, Atlantic Trade*, pp. 50–51.
11 P. D. Curtin, "The British sugar duties and West Indian prosperity", *The Journal of Economic History* 14 (1954), 157.
12 N. Deerr, *The History of Sugar*, vol. II (London: Chapman & Hall, 1950), p. 423.
13 Deerr, *History of Sugar*, vol. II, p. 532.
14 Bolland, *The Politics of Labour*, pp. 107–108.
15 Bolland, *The Politics of Labour*; Curtin, "British sugar duties", pp. 161–162.
16 Curtin, "British sugar duties", pp. 161–162; Bolland, *The Politics of Labour*, pp. 107–108; J. P. Greene, "Society and economy in the British Caribbean during the seventeenth and eighteenth centuries", *The American Historical Review* 79 (1974), 1499–1517.
17 P. Chalmin, *Tate and Lyle: The Making of a Sugar Giant, 1859–1989* (London: Routledge, 1990), p. 1; W. Kelleher Storey, *Science and Power in Colonial Mauritius* (Rochester: Rochester University Press, 1997), p. 39; *Report of the West India Royal Commission* (1898) C.8655.
18 Deerr, *History of Sugar*, vol. II, p. 531.
19 B. Richardson, "Depression riots and the calling of the 1897 West India Royal Commission", *New West Indian Guide* 66 (1992), 169–191.
20 *Report of the West India Royal Commission* (1898) C.8655.
21 H. A. Will, "Colonial policy and economic development in the British West Indies, 1895–1903", *The Economic History Review* 23 (1970), 129–147; S. B. Saul, "The economic significance of 'constructive imperialism'", *The Journal of Economic History* 17 (1957), 173–192.
22 Havinden and Meredith, *Colonialism and Development*, pp. 87–90; Saul, "The economic significance of 'constructive imperialism'", pp. 173–192.
23 Hodge, *Triumph of the Expert*, p. 60.
24 Kelleher Storey, *Science and Power*, pp. 98–100.
25 J. H. Galloway, "Botany in the service of empire: the Barbados cane-breeding program and the revival of the Caribbean sugar industry, 1880s-1930s", *Annals of the Association of American Geographers* 86 (1996), 682–706. The B prefix indicated a variety developed in Barbados while D stood for Demerara so indicated a variety developed in British Guiana.
26 Saul, "The economic significance of 'constructive imperialism'"; Havinden and Meredith, *Colonialism and Development*, pp. 87–90; Will, "Colonial policy and economic development".
27 Saul, "The economic significance of 'constructive imperialism'", p. 181.
28 Chalmin, *Tate and Lyle*.
29 *Report of the West Indian Sugar Commission* (1929), Cmd 3517.
30 *Report of the West Indian Sugar Commission* (1929), Cmd 3517.
31 *Report of the West India Royal Commission* (1945), Cmd 6607.
32 Deerr, *History of Sugar*, vol. II, p. 531.
33 See table in *Report of the West India Royal Commission* (1945), Cmd 6607, p. 25.
34 *Trinidad and Tobago Disturbances 1937: Report of the Commission* (HMSO, 1938); Havinden and Meredith, *Colonialism and Development*, p. 155, Table 7.3.
35 Library of the University of the West Indies, St Augustine Trinidad, "Caroni Ltd, Annual Report" (1975); Chalmin, *Tate and Lyle*.
36 *Report of the West India Royal Commission* (1945), Cmd 6607.
37 *Report of the West India Royal Commission* (1945), Cmd 6607.
38 *The Economist*, "The sugar problem" (29 March 1930), p. 694.
39 The National Archives, London (TNA), CO 295/606/4.
40 *Report of the West India Royal Commission* (1945), Cmd 6607.
41 TNA, CO 295/599/14.
42 *Report of the West India Royal Commission* (1945), Cmd 6607.
43 Barbados National Archives (BNA), GH4/111.
44 BNA, GH4/111.

45 S. Ryan, *Race and Nationalism in Trinidad and Tobago: A Study of Decolonization in a Multi-Racial Society* (Toronto: University of Toronto Press, 1972), pp. 44–47.
46 BNA, GH4/109.
47 W. A. Beckles, *The Barbados Disturbances, 1937, Review – Reproduction of the Evidence and Report of the Commission, Bridgetown* (Barbados: Advocate Co Ltd, 1937).
48 Beckles, *Barbados Disturbances*.
49 *Trinidad and Tobago Disturbances 1937: Report of the Commission* (HMSO, 1938).
50 Rhodes House Library, University of Oxford, J. Jagger, "Trinidad and the Tribunal, November 1938–February 1939. A Descriptive Diary", Saturday 21 January 1939, p. 53.
51 Jagger, "Trinidad and the Tribunal", p. 53.
52 M. Thomas, *Violence and Colonial Order: Police, Workers and Protest in the European Colonial Empires, 1918–1940* (Cambridge: Cambridge University Press, 2012), p. 224.
53 *The Times*, 5 May 1938.
54 *The Times*, 6 May 1938.
55 *The Times*, 10 May 1938.
56 Ashton and Stockwell, *Imperial Policy*, p. lxvi.
57 Havinden and Meredith, *Colonialism and Development*, pp. 195–205; Constantine, *British Colonial Development Policy*, ch. 7; Morgan, *Official History of Colonial Development*, vol. 1, ch. 4; Goldsworthy, *Colonial Issues*, p. 11; Lee and Petter, *The Colonial Office, War and Development Policy*.
58 Roger Louis, *Imperialism at Bay*, pp. 101–103.
59 H. Johnson, "The British Caribbean from demobilization to constitutional decolonization", in J. M. Brown and Wm. Roger Louis (eds), *The Oxford History of the British Empire*, Vol. VI, The Twentieth Century (Oxford: Oxford University Press, 2001), pp. 610–611; H. Johnson, "The West Indies and the conversion of the British official classes to the development idea", *The Journal of Commonwealth and Comparative Politics* 15 (1977), 55–83.
60 Constantine, *British Colonial Development Policy*, pp. 231–232.
61 Havinden and Meredith, *Colonialism and Development*, pp. 195–205; Constantine, *British Colonial Development Policy*, ch. 7; Morgan, *Official History of Colonial Development*, vol. 1, ch. 4; Goldsworthy, *Colonial Issues*, p. 11; Lee and Petter, *The Colonial Office, War and Development Policy*; Johnson, "The West Indies and the conversion of the British official classes".
62 *Report of the West India Royal Commission* (1945), Cmd 6607.
63 *The Times*, 3 July 1940.
64 TNA, CO 852/280/8.
65 Constantine, *British Colonial Development Policy*, p. 235.
66 It included the economist Hubert Henderson, Rachel Crowdy, Morgan Jones, Walter McLennan Citrine (General Secretary of the TUC), the agriculturalist Frank Engledow, Percy Graham Mackinnon, Ralph Assheton and Mary Georgina Blacklock, assisted by Thomas Lloyd and Charles Carstairs from the Colonial Office.
67 *Report of the West India Royal Commission* (1945), Cmd 6607, p. xiii.
68 Constantine, *British Colonial Development Policy*, ch. 9.
69 TNA, CO 847/15/9.
70 TNA, CO 847/15/9.
71 For an interesting consideration of the range of functions of Royal Commissions, see B. Lauriat, "'The examination of everything': Royal Commissions in British Legal History", *Statute Law Review* 31 (2010), 24–46.
72 Chalmin, *Tate and Lyle*, p. 220. Sugar rationing began on 8 January 1940.
73 TNA, CO 852/280/8.
74 *Report of the West India Royal Commission* (1945), Cmd 6607, p. 27.
75 *Report of the West India Royal Commission* (1945), Cmd 6607, p. 27.

76 TNA, CO 852/280/8.
77 TNA, CO 852/280/8.
78 W. Reader, *Imperial Chemical Industries: A History* (Oxford: Oxford University Press, 1970), pp. 322–323.
79 R. Weir, *The History of the Distillers Company: Diversification and Growth in Whiskey and Chemicals* (Oxford: Clarendon Press, 1995), p. 330.
80 TNA, CO 852/280/8.
81 *The Times*, 1 July 1919; *The Times* 28 January 1920.
82 H. Bernton, W. Korarik, and S. Sklar (eds), *The Forbidden Fuel: Power Alcohol in the Twentieth Century* (New York: B. Griffin, 1982), p. 27.
83 Bernton et al., *Forbidden Fuel*, pp. 15–23.
84 H. M. Petroleum Executive, *Report of the Inter-Departmental Committee on various matters concerning the production and utilization of alcohol for power and traction purposes* (1919), Cmd 218, p. 7.
85 *Ibid.*, p. 8.
86 K. Vernon, "Microbes at work: micro-organisms, the DSIR and industry in Britain, 1900–1936", *Annals of Science* 51 (1994), p. 605.
87 Vernon, "Microbes at work", p. 600; J. Reinharz, "Science in the service of politics: the case of Chaim Weizmann during the First World War", *The English Historical Review* 100 (1985), pp. 572–603.
88 Vernon, "Microbes at work", p. 600.
89 J. J. Hastings, "Development of the fermentation industries in Great Britain", *Advances in Applied Microbiology* 14 (1971), pp. 13, 24.
90 Hastings, "Development of the fermentation industries", p. 11.
91 Hastings, "Development of the fermentation industries", p. 14.
92 Fuel Research Board, *Fuel for Motor Transport* (HMSO, 1921).
93 Fuel Research Board, *Fuel for Motor Transport: An Interim Memorandum* (HMSO, 1920).
94 On the history of United Molasses, see, W. A. Meneight, *A History of the United Molasses Company Ltd* (Liverpool: Seel House Press,1977).
95 Weir, *History of the Distillers Company*, p. 298.
96 TNA, CO 852/280/8.
97 Weir, *History of the Distillers Company*, p. 288.
98 Weir, *History of the Distillers Company*, p. 286.
99 A. S. Travis, "Modernizing industrial organic chemistry: Great Britain between two world wars", in A. S. Travis, H. G. Schroter, E. Homburg and P. Morris (eds), *Determinants in the Evolution of the European Chemical Industry, 1900–1939* (Dordrecht: Springer, 1998), pp. 171–198. Here pp. 189–191.
100 L. F. Haber, *The Chemical Industry, 1900–1930: International Growth and Technological Change* (Oxford: Clarendon Press, 1971), pp. 90–92.
101 Haber, *Chemical Industry*, pp. 90–92.
102 TNA, CO 852/280/8.
103 *Ibid.*
104 *Ibid.*
105 *Ibid.*
106 NA, CO 852/280/8. The committee included Haworth, G. S. Whitby (Director of the DSIR Chemical Research Laboratory at Teddington), H. M. Bunbury (a Research Director of ICI), H. G. Paul (technical adviser to the Paper Controller of the Ministry of Supply), Philip Lyle and the Colonial Office's Agricultural Adviser, Harold Tempany.
107 TNA, CO 852/280/8.
108 *Ibid.*
109 TNA, CO 852/482/11.
110 TNA, CO 852/482/13.
111 TNA, CO 852/280/8.
112 TNA, CO 852/482/14.

113 TNA, CO 852/280/8.
114 J. Reinharz, "Science in the service of politics", pp. 572–603, Reinharz disputes the idea of a quid pro quo as too crude.
115 TNA, CO 852/482/14.
116 TNA, CO 852/482/16.
117 John Lionel Simonsen (1884–1957), assistant lecturer, Manchester University, 1907–10; Professor of Chemistry, the Presidency College, Madras, 1910–19; President, Chemistry Section, Indian Science Congress, 1917; Forest Chemist, Forest Research Institute and College, Dehra Dun, 1919–25; Honorary Secretary, Indian Science Congress, 1914–26; Professor of Organic Chemistry, Indian Institute of Science, Bangalore, 1925–27; Professor of Chemistry University of College of North Wales, Bangor, 1930–42; Vice-President and President, Chemical Society, 1940–55; Secretary, Chemical Society, 1945–49; member, ARC, 1944–49; President, Section B, British Association, 1947; Chairman of the Board, Pest Infestation Laboratory, 1949–52.

CHAPTER TWO

Scientific research and colonial development after 1940

In 1941 the Colonial Office made a commitment to fund scientific research into the chemistry of sugar. If sugar cane could be used to make plastics, building materials, drugs and other synthetic products, then it was hoped the British West Indies would find themselves in the fortunate position of being producers of a lucrative raw material for the chemical industry rather than a low-value foodstuff. This was a vision that endowed laboratory research with the power to transform the economic and social life of the British West Indies. But how exactly was knowledge expected to move from the laboratory and spur development? This chapter will examine the relationship between scientific investigation and colonial development that was embodied in the new arrangements for colonial research that were created in fields such as sugar chemistry during the first half of the 1940s.

The late colonial period saw an unprecedented expansion in scientific research across the Colonial Empire and in British universities, funded through the Research Fund of the 1940 CDW Act and its successors. The new research fund formed part of the Colonial Office response to the crisis that affected the Colonial Empire in the late 1930s, of which riots in the Caribbean were only a part. Research became a priority at a point at which Britain needed a meaningful gesture to ward off domestic and international criticism of the management of its colonies. Scientific research was described as a practical tool that would provide the basic information that underpinned development and so would serve to guarantee the efficacy of Colonial Office interventions in the future.

The Colonial Office engaged a group of high-powered scientific advisors drawn from Britain's research councils to oversee spending of funds from the CDW Acts. These scientists created a system for colonial research that included over forty research institutions in the colonies by 1952. The expression used most frequently in the early 1940s to describe the work that the Colonial Office wished to support in these institutions was 'fundamental research', and officials and scientists took great pains to differentiate this work from other modes of science such as problem solving and routine testing. The key value that informed the new arrangements was 'freedom'. Advisors at the Colonial Office claimed that for the highest quality research to occur, scientists had to be free to choose their own research problems and be free of direction by administrators and less-qualified technical officers. The arrangements that were introduced for colonial research were, in fact, less the expression of a rational model of the development process and more the product of the priorities of a group of elite metropolitan scientists who wished to provide autonomy and status for scientific researchers working in government service. The fact that the discourse on science and development that emerged in the 1940s could encompass both the idea of research as the basis of planning and research as an activity in which freedom for researchers was paramount was possible because of the multiple meanings that could be attached to the idea of fundamental research. This was a concept of considerable political utility.

Research and colonial development after 1940

In 1938, the freshwater biologist E. B. Worthington published a summary of the state of scientific research in Africa in which he made the following complaint about past development initiatives:

> Economic development has taken the lead and often chooses the wrong turning. Science follows, but the pace is laboured, and falling behind she is neglected. Often she has to follow along the wrong path for some distance before beckoning development back to the direct way. Roads and rails have been built before there were accurate maps on which to mark them; crops have been introduced and grown under all kinds of conditions, regardless of the suitability of the soil.

He concluded that, 'A development based on a real understanding of Africa's potentialities has hardly yet begun, and will be impossible until the necessity of scientific knowledge is recognized'.[1]

The idea that the failure of previous development projects was due to a paucity of knowledge about tropical conditions had great resonance. At the Colonial Office, a wider reform of colonial policy was under way that aimed to address the limitations of the 1929 Colonial

Development Act. The priority of the 1929 Act had been to alleviate unemployment at home by generating demand for British manufactured goods, and the restricted nature of loans from the Act had led to few improvements in social or economic conditions in the colonies. The deprivations experienced by many territories during the Great Depression, along with increasing hostility to British imperialism in the US and Germany, contributed to an acute sense of crisis amongst colonial officials by the end of the 1930s. The possibility that Britain could be made to relinquish its colonies altogether gave urgency to the idea that a grand gesture was required to show Britain was committed to taking action to deal with colonial problems, even when war had recently broken out. The CDW Act was formulated to provide free grants for development in both the economic and social sphere and to create a large fund solely dedicated to scientific research. For the Colonial Office the timing was crucial, 'Measures for the advancement of the Colonies are politic at a time when the general question of Colonial responsibilities is under widespread criticism and when it is expedient for us to justify our position.'[2]

Interest at the Colonial Office in an expansion of colonial research had in fact existed for some time. While the rise of development as a goal of colonial policy from the 1890s onwards was accompanied by a growing belief in the importance of science and medicine, funds specifically for research were not plentiful before the Second World War. Small ad hoc grants for research were issued by the Colonial Office between 1919 and 1933, and the Empire Marketing Board also allocated research monies of £285,000 in the period between 1926 and 1933. A small number of grants for research came from the 1929 Colonial Development Fund. In the field of agricultural research, the Lovat Committee made recommendations for an expansion of agricultural research in 1927 and this helped to revive the fortunes of the East African Agricultural Research Institute at Amani in Tanganyika. This research station, created by Germany when it had been in possession of Tanganyika, had fallen into dereliction, and the neglect of Amani was presented as symbolic of the relative indifference of Britain to research.[3] Whilst funds were secured for Amani, and also for the Imperial College of Tropical Agriculture in Trinidad, the wider proposals of the Lovat Committee for a chain of agricultural research stations were never implemented because of the refusal by colonial governments to provide financial support and the retrenchment of the 1930s.[4] Research in medicine and social science in the colonies was mainly funded by the Rockefeller Foundation before 1940 with addition of some small grants from the MRC.[5] Worthington's *Science in Africa* noted some of the gains that had been made by research in Africa, including studies of rinderpest, east coast fever and other diseases of cattle, sheep and goats in Kenya and plant-breeding programmes that had increased the yields of sorghum in Nigeria.[6] Research, however, was an activity that members of medical, veterinary and agricultural departments in the colonies often had to fit in around their regular duties. In general, while the early twentieth century saw a substantial increase in the numbers of scientists and medical personnel deployed to work in the colonies, research work was often marked by a lack of continuity and coordination. Officials in London and colonial governments were not in direct command of substantial funds or a cadre of scientists to implement programmes of investigation.[7]

Colonial Office plans for the creation of a specially designated research fund as part of the 1940 CDW Act drew upon the recommendations of the eminent figure of Malcolm Hailey. Lord Hailey had headed the African Research Survey that had recruited Worthington to undertake his study of scientific provision in Africa. Hailey had agreed to lead the Carnegie-funded African Research Survey after an illustrious career in India that culminated in the post of Governor of the United Provinces. The African Research Survey of 1936 was an ambitious attempt to capture both the present state of knowledge about Africa and the potential for its future development based on scientific understandings. It had its origins in meetings at Oxford University in 1929 that brought together African experts, politicians, high-ranking civil servants and scientists, including Leopold Amery, Frederick Lugard, Joseph Oldham, Malcolm MacDonald and Julian Huxley.[8] As a result of his tour of Africa, Hailey produced *An African Survey*, a landmark account of problems affecting the African colonies of the European powers and the paths to future progress across a large number of fields such as science, law and anthropology. *An African Survey* generated a great deal of public attention and officials at the Colonial Office would come to refer to it as 'the bible on practically everything relating to our administration in tropical Africa'.[9] This substantial volume incorporated Worthington's recommendations for a major expansion in research. Hailey stated that development could never be accomplished while the colonial powers were ignorant of the basic conditions that existed across the African continent. The priority was, 'a more comprehensive study of the factors which determine the nature of its social development, and a more scientific approach to the problems of health and material well-being to which its physical characteristics have given rise'.[10]

The Secretary of State for the Colonies, Malcolm MacDonald, was persuaded to include a research allocation extended to the study of issues occurring across the whole of the Colonial Empire as part of the proposed CDW Act. With the passing of the 1940 CDW Act this was

£500,000 each year, doubling to £1 million with renewal of the Act in 1945. The Research Fund was Britain's most significant financial commitment to research related to the problems of the British Colonial Empire by a considerable margin. It elevated the Colonial Office to the position of second-largest sponsor of civil scientific research in Britain during the 1940s. Suddenly the Colonial Office had funds that eclipsed the allocation given to the ARC and the MRC, with only the DSIR, a government department entirely devoted to promoting scientific research, receiving a larger provision.[11] The Research Fund, then, was a major new source of support for colonial research, and also for British scientists.[12]

Alongside the new fund, Hailey persuaded Malcolm MacDonald to appoint a body at the Colonial Office of leading British scientists to oversee the spending of the money.[13] The stature of the proposed committee was important: 'Apart from being competent to give advice on research schemes, the committee should be such as to command the respect and confidence of the Colonial Office Advisory Committees and the scientific world in general.'[14] Hailey made it clear in *An African Survey* that the success of any expansion in colonial research would be dependent upon recruiting suitably well-qualified personnel to undertake these investigations.[15] There was a feeling, however, amongst those that were familiar with the scientific and medical work prosecuted in the colonies before 1940 that the technical services – the Colonial Medical Service, the Colonial Agricultural Service and the Colonial Veterinary Service – generally failed to attract talented researchers. These services employed two types of candidate. Some officers in the Colonial Agricultural Service, for example, came with a general qualification in agriculture that made them well suited to the task of teaching farmers in the field. These members of departments of agriculture in the colonies were often confusingly referred to as 'administrators'. Other officers had degrees in what G. B. Masefield, the historian of the Colonial Agricultural Service, described as 'pure science'.[16] This was a reference to men who had specialisms in fields such as zoology or plant physiology rather than something more practical such as animal husbandry, and who could undertake experimental study.[17] Efforts had been made to raise the standard of both types of candidate during the interwar period. New advisory committees were formed at the Colonial Office to oversee and coordinate work in areas such as health and agriculture. A Chief Medical Adviser was appointed in 1926 and an Agricultural Adviser in 1929, and this helped to raise the profile of technical matters in London. In 1934 and 1935 the Colonial Office unified the regional branches of the colonial services that employed staff for agriculture, veterinary medicine, forestry and medicine with the intention of giving better career prospects for officers as they could now move to positions across the whole Colonial Empire. To further raise the prestige of the services, entrants were increasingly given specialised training; cadets for the Colonial Agricultural Services took a probationary course at the University of Cambridge and the Imperial College of Tropical Agriculture in Trinidad from 1926. The view expressed in 1940, however, was that the numbers of individuals suited to research were still low. Thomas Lloyd, Assistant Under-Secretary of State at the Colonial Office, stated that while training had worked to raise the level of administrative officers, the colonial services were not successfully competing with domestic research institutions for specialist scientific staff.[18]

The idea that the colonial services failed to attract really good research workers lay behind Hailey's suggestion of a committee that would raise the stature of colonial research. The goal was to place the high-profile committee at the Colonial Office on the same footing as the research councils in Britain. In 1940 there were three research councils that provided money for science in British universities and in their own research institutes. The DSIR had been created in 1916, followed by the MRC in 1920 and the ARC in 1931. The research councils were not the only source of funds for British scientists during the interwar period. Government departments such as the Ministry of Health or the Ministry of Agriculture and Fisheries also funded science, and the Admiralty, Air Ministry and War Office spent substantially more on research than the largest research council, the DSIR.[19] The research councils, however, claimed a special position in British science on the basis that, unlike government departments, they were not under the control of a minister or a senior member of the armed services, but instead were overseen by a committee of scientists who reported to the Lord President of the Privy Council. Edward Mellanby, Secretary of the MRC in 1940, claimed that the significance of this arrangement lay in the fact that the MRC was not subject to direction by a non-scientific administrative class. Scientists determined the policies of the MRC and therefore the direction of medical research in Britain. Accordingly, it was claimed, the MRC was a body that was in receipt of state funds but was not subject to state control.[20] Mellanby asserted that it was this position of independence that gave the MRC its special status in the eyes of British medical researchers, and it was clear that for many the research councils were considered to be the true bastions of fundamental research in Britain.[21]

The problem facing the Colonial Office as it contemplated a significant expansion in colonial research was the relative disdain that could be shown for the science prosecuted by government departments, at home and in the colonies.[22] Hailey and the Colonial Office believed the high-calibre scientific researchers they sought were not likely to seek employment in the colonial services unless a system for research in the colonies was created that had clear connections with the DSIR, ARC and MRC. The first step was the creation of a committee, the Colonial Research Committee (CRC), along the lines suggested by Hailey and which contained eminent scientists such as Mellanby recruited from the research councils. In 1945, the Colonial Office also created a new internal department in the Economics Division dedicated to the administration of colonial research, headed by an administrator, Charles Carstairs. The Research Department was described by officials as a 'Department of Scientific and Industrial Research to the Colonial Office'; clear indication that the Colonial Office wished to see its work in the field of colonial research aligned with the elite system to administer state funds for research that had developed in Britain since the First World War.[23] The suggestion by Hailey and Worthington that research would provide the knowledge that would underpin effective development struck a particular chord with Carstairs, who stated in 1945:

> I think that it must be recognised that in the prevailing absence of the bulk of the fundamental data required for sound planning, much of the developmental expenditure cannot fail to be misdirected and so wasted, together with the man-power diverted thereto. We must reconcile ourselves to a period of building on sand, and to some extent of pouring money into sand, but we should I think make it a primary object of policy probably for the duration of the CD and W legislation to reduce this period of inevitable waste and disappointment by making a very serious effort to construct a solid framework of basic information by means of the survey techniques listed earlier.[24]

Later, Carstairs circulated to his colleagues an address by Herbert Morrison in which Morrison had described five stages in planning. The second stage in this process, Carstairs noted, was the collection of data.[25]

Colonial Office enthusiasm for a research fund and new research arrangements as part of the wider reform of policy that cumulated in the 1940 CDW Act can be explained by a desire on the part of officials to see colonial development be placed on a much firmer footing. The claim that research would provide the knowledge that ensured the efficacy of development had great appeal at a moment in which Britain was sensitive to criticism about poor management of its colonial possessions. In their application to the Treasury for the creation of the Research Fund, officials spoke of the need to 'substantiate, in as striking manner as possible, its professions of trusteeship on behalf of the subject peoples in the colonial dependencies'.[26] The significance of the discussions that led up to the CDW Act and its Research Fund lay not just in the idea that greater funds be created, however, but also in the suggestion that scientific research needed new arrangements to recruit the personnel who were going to execute a comprehensive programme of investigation into the problems affecting the Colonial Empire.

New arrangements for colonial research

The CRC had its first meeting in June 1942, and at that time the committee included the heads of the research councils – Edward Mellanby[27] (MRC), W. W. C. Topley (ARC)[28] and Edward Appleton[29] (DSIR), along with A. V. Hill,[30] Secretary of the Royal Society. The scientists on the committee had recently met together as the Scientific Advisory Committee to the War Cabinet, under Lord Hankey.[31] They were a powerful and influential group who controlled the elite bodies for civil research in Britain and who were experienced in providing advice to government.[32] According to the scientist and administrator Alexander King,[33]

> Up to the Second World War, the size of the British science system was small enough for internal adjustments and policy direction to be in the hands of a few, outstanding personalities belonging to the same coterie. Coherence and mutual understanding were probably achieved rather effectively, if utterly informally, through frequent, easy, but often unplanned contacts between the leading figures of the Royal Society, the research council secretaries, and senior civil servants, all of whom were habitués of the Athenaeum Club.[34]

Hailey was appointed chair of the committee, and in 1946 Worthington was appointed Scientific Secretary. By the end of the 1942, a number of prominent representatives of business, social science and economics had also been invited to take a seat on the CRC: A. M. Carr-Saunders, Director of the London School of Economics, the social scientist Audrey Richards, the economists Henry Clay and Hubert Henderson, and John Caulcutt, Chairman of Barclays Bank.[35]

In addition to the CRC, a number of other committees were created between 1943 and 1947 that brought even greater numbers of eminent scientists linked to existing British research organisations, often through the research councils and the universities, into the Colonial Office. The Colonial Office formed the CPRC in 1943,[36] followed by the Colonial Fisheries Advisory Committee. The Colonial Social

Science Research Council and the Tsetse Fly and Trypanosomiasis Research Committee were formed in 1944 and the Colonial Medical Research Committee and the Committee for Colonial Agriculture, Animal Health and Forestry Research were formed in 1945. These committees were followed by the Anti-Locust Research Centre Scientific Committee in 1946, and the Colonial Economic Research Committee and the Colonial Insecticides Committee in 1947. The CRC had an overview of all research, ensuring, for example, the fair division of funds between different research areas, and it oversaw work in a number of miscellaneous areas such as buildings research.

Even though positions on the research committees were voluntary and unpaid, the Colonial Office gave these bodies considerable power; they were free to devise their own research projects and they decided the future of all schemes submitted to the Colonial Office by colonial governments. During the 1940s, officials at the Colonial Office did not generally interfere in the business of the new research committees and often privileged the views of these research bodies over those of existing officers in the technical services in the colonies. In effect, responsibility in determining the future direction of colonial research was shifted to metropolitan committees that contained scientists who did not necessarily have any direct experience of work in the tropics but who had made careers in the domestic research system. The aim in doing this was to bring colonial research into much closer contact with research institutions and universities at home. This was considered essential to facilitate the recruitment of high-calibre scientists to work on colonial problems. One of the deterrents to working in the colonial services was said to be the fact that it distanced young scientists from the major centres of their disciplines so they struggled to keep abreast of developments in their field, and this retarded their careers in the long term.[37] The committees that oversaw research contained scientists who had contact with the main bodies and institutions for research in Britain and this was said to ensure better communication between science in Britain and its colonies. In addition, the high profile and reputation of members of the CRC was thought to endow colonial research with greater cachet than it would otherwise be able to attain.

The dominance of the new committees at the Colonial Office by individuals linked to metropolitan centres of research, and the considerable authority given to these committees, proved significant for the organisation of colonial research in a number of ways. The scientists drawn from the domestic research councils that sat on the Colonial Office committees impressed upon officials that colonial development required an expansion in fundamental research. Long-term fundamental research was said to provide the knowledge of soils, insects, crops, disease and commodities on which all other activities, from problem solving in science to efficient management of the development process, ultimately depended. It was said to yield an in-depth understanding of the most widespread and basic issues that existed in the colonies so that fundamental research was in fact 'general' research.[38] Importantly, this was an assertion about the nature of research only. Other activities that fell within the scope of science, such as laboratory testing, the preparation of vaccines or the short-term solution of practical problems, were intrinsically 'local' problems relating to the immediate requirements of the individual territory, and were best dealt with by the existing technical departments in the colonies. The investigation of fundamental problems that were shared by colonies, on the other hand, was said to require the view from above provided by the London-based research committees.

The creation of the CRC and other research committees worked to marginalise the input of scientists, veterinary officers and doctors that were already based in the colonies into shaping both new arrangements for research and even its content. Some doctors, biomedical scientists and veterinary officers had sought to coordinate their work in the colonies on a regional basis and expand their research activities before the outbreak of the Second World War. The CRC explained why colonial research was an activity that required new arrangements, however, that shifted control of research to committees based in London. In the colonies, 'there is a tendency for research problems to be dictated too exclusively by local and temporary interests, without due regard to scientific possibilities, or to the scale on which a given investigation must be planned if it is to have any reasonable hope of success'.[39] And 'the frontiers of scientific research do not coincide with political boundaries. In so far as scientific problems in various parts of the world resemble one another, the boundaries are rather lines of latitude.'[40]

Research considered problems that eclipsed the borders of individual colonies and therefore the organisation of this work was beyond the scope of the technical departments that operated as part of the colonial governments. Only committees such as the CRC at the Colonial Office in London were able to grasp the scale of the problems dealt with by research.

References to the nature of research had rhetorical intent; they worked to naturalise the ambitions and preferences of elite scientific advisors to the Colonial Office. In part, declarations about the intrinsic character of research worked to assure the position of the research councils and their representatives in controlling the colonial research agenda after 1940. This concern for their own status was characteristic of bodies such as the MRC. The system of state funds for science in Britain had grown considerably since the First World War. From discussions concerning the creation of the DSIR onwards, British scientists had worked hard to negotiate between a desire on the one hand to receive greater funds for the work they wished to undertake and the need to reject any possibility that government funding would mean civil servants or politicians would decide the goal of their investigations. For many scientists, state funding of science raised the spectre of a shift towards applied science in which administrators would determine the objectives of research. The promotion of fundamental research as an activity in which oversight by scientists was essential was one way in which researchers attempted to maintain their professional standing. Fundamental research was synonymous with the preservation of freedom for research workers to pursue their investigations along lines of their own choosing. The fear that a closer relationship with government could result in the subordination of scientists to the interests of the state was also apparent in the debates about the position of scientists in the civil service. The 1943 Barlow Report created the Scientific Civil Service, in which scientists were placed for the first time on grades that were equivalent in pay and status to their non-scientific counterparts. Apart from addressing the grievances of many scientists, Barlow's recommendations were intended to improve the image of the civil service as a career for talented researchers.[41] The discussions at the Colonial Office about the appropriate machinery for the organisation of colonial research occurred therefore against a wider backdrop in which British scientists were keen to improve their status in relation to administrators in government and in which the research councils had emerged as strident defenders of autonomy for scientific researchers in order to preserve their status.

Meetings of the CRC show it to be preoccupied by the relative status of research staff engaged to work on colonial problems. The CRC said of the highly qualified research staff it hoped to recruit for work on colonial problems,

> workers must be not be [sic] inhibited from working in the way best calculated to allow them to achieve the most valuable results, which means, in the field of scientific research as much as in any other sphere of creative activity, allowing the worker the greatest possible latitude as to his methods of work. Complete freedom of enquiry is not the only, but it is an essential, condition of fruitful research work.[42]

Freedom of enquiry could only be guaranteed with the right people in charge. This was not just an argument for ensuring that non-scientific administrators did not tell scientists what projects they should work on, it was an argument that other types of technical and scientific officer would not supervise research staff. In the Agricultural Department this might mean making sure that an extension officer with a degree in agriculture was not in charge of a scientist with a specialist degree from Cambridge University. Worthington had argued in *Science in Africa*, and the CRC agreed, that an expansion in research activities related to the needs of colonial development could not be achieved by merely giving more funds to the technical departments of the colonial governments. Research needed to be separated from other scientific and medical work and placed under the control of appropriately qualified research staff. It was said that the most able research workers simply would not submit to the authority of agriculturalists (called 'administrators' in the Colonial Agricultural Service) who lacked research experience. To lend weight to this argument, researchers were differentiated from other technical staff on the basis of their temperament: 'We doubt whether anyone who has not been an active research worker for a part of his life can effectively lead a research team with the understanding and appreciation that will bring out the best of which members are capable.'[43] And 'The normal administrator, with ideas based on command and orderliness, must find it difficult to accept a position as a leader of a group of individualists; and attempts to impose discipline or order on the research worker can be fatal to productive research.'[44]

The need to ensure freedom for scientists working on research, the separation of research from other scientific and medical work in the colonies, and the focus on problems that transcended the boundaries of individual colonies, provided the rationale for the creation of a range of new institutions in the Colonial Empire.[45] Forty new institutions and research units were established between the Second World War and 1952, many of which operated on a regional basis (see Table 3). The majority of the new research units were built in Africa, accounting for the larger proportion of the Research Fund spent on this area of the Colonial Empire (East Africa alone was given an allocation of 39 per cent of the total). The largest proportion of research monies (36 per cent) was spent on research in agriculture, animal health and forestry, reflecting the central importance given to these fields in the economic life of the British Colonial Empire.[46] Two of the new laboratories were created under the aegis of the CPRC, the Sugar Technology Research Unit and the Colonial Microbiological Research Institute in Trinidad.

The new research units were afforded considerable autonomy with respect to colonial administrations, and research staff were not required to incorporate the suggestions of Directors of Agriculture or Medicine in the colonies in the development of their projects.

Table 3 Research institutions in Britain's colonies funded by the CDW Acts, 1940–52.

Region	Name of institute
East Africa	Fisheries Research Institute (Jinja, Uganda)
	East African Institute of Social Research (Makerere College, Uganda)
	Cotton Research Station (Uganda)
	Virus Research Institute (Entebbe, Uganda)
	East African Insecticides Research Unit (Arusha, Tanganyika)
	East African Agricultural and Forestry Research Organisation (Kenya)
	East African Veterinary Research Organisation (Kenya)
	East African Tsetse and Trypanosomiasis Research and Reclamation Organisation
	Filariasis Research Unit (East Africa)
	East African Scientific and Industrial Research Organisation
	East African Malarial Unit
	Coffee Research Station (Kenya)
West Africa	Virus Research Institute (Lagos, Nigeria)
	West African Institute for Tsetse Fly and Trypanosomiasis Research
	Nutrition Field Research Station (Gambia)
	West African Fisheries Research Institute
	Sir Alfred Jones Laboratory (Sierra Leone)
	West African Veterinary Research Organisation
	Rice Research Station (Rokupr, Sierra Leone)
	West African Institute of Social and Economic Research
	West African Road Research Laboratory
Central Africa	Rhodes-Livingstone Institute (Northern Rhodesia)
	Agricultural Research and Experimental Station (Nyasaland)
	Tsetse Fly Research Unit (Northern Rhodesia)
	Fisheries Research Organisation (Northern Rhodesia)
Caribbean	Low Temperature Research Station (Trinidad)
	Colonial Microbiological Research Institute (Trinidad)
	British West Indies Sugar Cane Breeding Station (Barbados)
	Ebini Livestock Experimental Station (British Guiana)
	Sugar Technology Laboratory (Trinidad)
Asia	Timber Research Station (Malaya)
	Institute for Training in Fish Farming (Penang Island, Malaya)
	Fisheries Research Station, (Hong Kong)
Pacific	Institute of Educational Research (Fiji)

Recipients of grants from the Research Fund communicated with the London-based committees about the work they did and it was the opinion of these bodies that mattered. This separation of research work from other types of medical and technical work was in many ways the reproduction of the system that had emerged in Britain in which the research councils were the sole determinants of the work done in the institutes they funded. Government departments did not set out the objectives of research done through the research council system, so that the Ministry of Health did not decide the subjects to be investigated by the MRC. Similarly, technical departments in the colonial governments did not determine all the research that happened in their midst.

The circumvention of local authority, and the central role given to the research committees at the Colonial Office when it came to the oversight of research, was intended to enhance the status of these London-based committees and the research staff they appointed to work on colonial problems. However, the semi-autonomous position of new research institutions in the colonies raised the issue of communication between technical departments and research workers. In some cases, new regional advisory committees were formed to enable some liaison between departments of agriculture and agricultural research institutes, for example. Scientific advisors at the Colonial Office were clear, however, that these arrangements did not mean that Directors of Research at East African research organisations would be submitting to the authority of the Director of Agriculture of the Kenyan administration, for example.[47]

The Colonial Office firmly endorsed the new arrangements, explaining the principle of freedom for researchers and the special position of autonomy given to the CRC and other bodies that oversaw colonial research to African governors in 1947.[48] Officials believed that the obstacle to a significant expansion in the amount of research carried out across the Colonial Empire was the fact that government departments and government service generally failed to attract the most well-qualified and research-orientated science graduates. The provision of working conditions that the placed the control of research firmly in the hands of established research workers and the clear alignment of colonial research with the work of the domestic research councils were considered necessary in order to confer upon colonial research the prestige required to attract high-flying scientists to work on colonial problems.

A willingness by officials to see the administration of colonial research placed under the control of metropolitan committees also reflected a wider ethos at the Colonial Office that valued innovation by metropolitan experts. The passing of the 1940 CDW Act, followed by the experiences of wartime mobilisation, led to a shift to a more assertive, interventionist and centralised approach by the Colonial Office to the pursuit of economic and social development in Britain's colonies.[49] Sydney Caine expressed his concern in a famous Colonial Office memorandum of 1943 that colonial governments were failing to submit sufficiently ambitious and well-articulated plans for development. Speaking about economic planning, Caine stated that the problem was that the colonial administrations did not have the necessary expertise to devise plans that would produce real change in the colonies. He urged the Colonial Office to move from its traditional role of merely assessing submissions as they came in from the colonies and seize the initiative in devising solutions to colonial problems. Caine expressed his approval of the CRC as a body that did not passively wait for colonial governments to submit schemes for approval but was active in producing plans for research and the new machinery needed for its prosecution.[50]

Research into colonial products after 1943

Tropical products were subject to technical investigation by the state and business from the nineteenth century onwards but the creation of the CPRC in 1943 marked the beginning of a new episode. In the past, bodies such as the Imperial Institute had analysed and assessed commodities as part of a commercial intelligence service. In contrast, the CPRC promoted the fundamental chemical investigation of tropical products. This was a shift to long-term and exploratory laboratory research of the chemical constituents and synthetic pathways of natural products, the sort of in-depth investigation of tropical conditions and materials that many scientists had argued should be prioritised with the creation of the Research Fund of the 1940 CDW Act. The arrangements that were put in place for this research reflected many of the claims about the special conditions needed to facilitate fundamental research made by the CRC and other bodies of scientific advisors at the Colonial Office. The work of the CPRC in London contributed to the centralisation of colonial research, with control over decisions about the fields of scientific enquiry that would be pursued residing with metropolitan scientists, and researchers receiving funds from the CPRC given a great deal of freedom in their work. Colonial governments, technical officers in the colonies and businessmen made little contribution to the nomination of research problems. The idea of fundamental research as the investigation of the basic and general issues relating to the chemistry of natural products was also important for the negotiation of the relationship between state funds and commercial interests in the vision of development promoted after 1940.

Before 1940 the evaluation and improvement of plant and animal products from the colonies had been one of the functions of the Imperial Institute. The institute opened in 1893 at South Kensington in London as part of a complex of buildings created to celebrate the Golden Jubilee of Queen Victoria. It had a number of functions that amounted to the promotion of empire to the British public and British business. The Scientific and Technical Research Department evaluated colonial products such as rubber, cotton, medicinal plants, tanning agents, fibres, oil nuts and minerals submitted by the colonies with the aim of encouraging greater utilisation of empire products by metropolitan firms.[51] The value of this work was brought into question, however, as the colonies expanded their own capacity for the analysis of tropical products. British chemists were recruited in increasingly large numbers in the first half of the twentieth century for work in the Colonial Chemical Service and other branches of the colonial service.[52] A further push to employ state-funded science in the greater exploitation of empire products occurred in 1926 with the creation of the Empire Marketing Board. The EMB issued grants totalling around £3.5 million to universities and research institutions with the goal of improving the quality of tropical products so that commodities from the British Empire could compete with those from other sources.[53] The EMB had a political function in promoting the domestic consumption of empire goods without recourse to the protectionist measures that were so unpopular with the public (the prospect of tariffs and increased food prices had led to the Conservatives losing their overall majority in the election of 1923).[54] The decision to introduce imperial preference at the Ottawa conference in 1931 removed the original rationale for the existence of the EMB, however, and it was disbanded in 1933.

Business also took measures to evaluate and improve tropical commodities, not least in response to the threat to markets represented by the emergence of synthetic alternatives. Indigo planters in Bihar created new laboratories between 1898 and 1905 for chemical investigations to improve the quality and yield of the natural dye in the face of declining demand for their product with the development of synthetic indigo in 1897.[55] In addition, firms carried out investigations intended to expand and improve their range of products. The Niger Company analysed vegetable oils and rubber in its London Produce and Development Department and undertook surveys in Nigeria to search for new mineral, botanical and animal products.[56] By the Second World War, experimental stations to improve cash crops such as coffee, cocoa and rubber had been established in a number of colonies, funded in part or wholly by business. In 1902 the British Cotton Growing Association was formed and established research stations in Africa for breeding disease-resistant strains of cotton.[57]

The role of the CPRC differed from these earlier attempts to aid the exploitation of natural products from Britain's colonies. Rather than providing a service that assessed the quality of colonial products, or using plant-breeding techniques to improve them, the council sponsored laboratory research to identify new ways that existing products could be utilised. In the words of the Secretary of State for the Colonies, Oliver Stanley, in 1943 the goal of the CPRC was 'to save the old products by finding new uses for them'.[58] Investigations focused on reinventing tropical products as the raw materials for manufacturing synthetic goods at a point at which chemical firms in Europe and America were poised to produce increasingly large numbers of plastics, synthetic fabrics and therapeutic materials through organic chemistry. The focus of the CPRC was on the long-term and in-depth investigation of the many different chemical pathways and transformations of a substance like sucrose in order to take advantage of 'the brave new synthetic world that is growing up around us', in the words of the *Manchester Guardian*.[59]

Fundamental research into the chemistry of natural products was considered a new area for the Colonial Office, and as with other fields of colonial research, it was considered necessary to form a committee of leading scientists drawn from British universities and research establishments in order to oversee spending from the Research Fund. The secretaries of the DSIR, MRC and ARC all sat on the CPRC and they were joined by a number of eminent chemists – Professor Norman Haworth, who headed a team of carbohydrate chemists at the University of Birmingham, Robert Robinson, Head of the Dyson Perrins Laboratory at Oxford, the Government Chemist Sir John Fox, Sir John Simonsen, appointed Director of Research of the CPRC, and Professor Ian Heilbron of Imperial College of Science and Technology. Two members of the committee, Haworth and Robinson, had been awarded the Nobel Prize by 1947. To further raise its profile, the CPRC was chaired by the high-ranking official Maurice Hankey. In 1938 Hankey had formally retired from an extraordinary career that had begun in naval intelligence and included the post of Secretary of the Committee of Imperial Defence for twenty-six years, Clerk to the Privy Council, and Cabinet Secretary. Hankey had taken an interest in scientific and technical issues over his career, often in relation to defence matters. Most recently he had contributed to the Barlow Report on improved pay and status for scientists in the civil service and he was appointed the Chair of the Scientific Advisory Committee to the War Cabinet in 1940. This committee aimed to coordinate the wartime research efforts that were happening across government. Hankey combined a role as chair of the CPRC with a leading position in the post-war organisation of Britain's programme of biological warfare research. In 1946 he was appointed Chair of the Bacteriological Research Advisory Board that created the Microbiological Research Establishment at Porton Down.[60] The CPRC also included Gerald Clauson, who was head of the Economics Division of the Colonial Office, G. W. Thomson of the Trades Union Congress, Aneurin Davis of the Co-operative Wholesale Society, and Harry Lindsay, who was Director of the Imperial Institute.

At its first meeting the CPRC created a technical sub-committee from the chemists on the council to determine its research programme. This committee of Haworth, Robinson, Simonsen, Fox and Heilbron chose the commodities to be investigated by the council and then either undertook this research themselves or secured the services of chemists working at British universities to undertake the relevant investigations. The procedures of the DSIR were used as a guide for determining the salaries and contracts that the CPRC offered scientists receiving its funds.[61] The DSIR was considered an appropriate model for the CPRC as its function was to encourage research in British universities and other laboratories related to the needs of British industry. Private business was expected to play a major part in fulfilling the ambitions of the CPRC, but no representative of a firm or a colonial government sat on the council, and business and technical staff in the colonies were not the main source of proposals for research. Instead, the scientists on the technical committee determined the scope of research undertaken by the CPRC and placed control over the direction of a research project in the hands of the scientists they sponsored.

Sugar remained the commodity of central interest.[62] In addition to the production of alcohol for fuel, the committee focused on using sugar as a starting material for making solvents, plastics, drugs and anti-freeze agents. At the first meeting of the CPRC in January 1943 Haworth reiterated his claim that sugar was a cheaper starting material for the expanding organic chemical industry than oil or coal.[63] Coal had been the raw material used by the German firm IG Farben and the American company Du Pont before the war, but Robinson stated to the CPRC that now, 'Coal was on its deathbed except as a fuel', confirming the need to explore alternative raw materials.[64] Haworth's claim of the potential of sugar as the starting compound for a range of synthetic chemical products was given credence by the news that some chemical derivatives of sugar were already under investigation by ICI. The chemical firm had contacts on the CPRC as it had sponsored research by Haworth and his colleague, Leslie Wiggins, at

Birmingham University in the past. In addition, Robinson had joined a research council created by ICI in 1927 where he made an important contribution to the development of one of the early plastics, polyethylene.[65] News came in 1943 that ICI was investigating pathways from sugar to manufacture the versatile chemical intermediate furfural. Furfural was used for making lubricating oils and thermosetting plastics, and an industry producing this chemical was well established in the US, operated by the Quaker Oats Company. A search for cheap and accessible sources of furfural became an urgent matter during the Second World War. Furfural was key to the manufacture of synthetic rubber, a compound in high demand as natural rubber imported from the tropics became increasingly difficult to secure with Japanese control of Malaysia.[66] Organic intermediates such furfural and levulinic acid were compounds of enormous utility, important for the manufacture of a huge range of industrial products. There was great demand for such intermediates by the chemical industry and therefore a lucrative opportunity appeared to exist if new and better chemical processes could be developed to produce such materials from sugar.

A number of meetings took place in 1943 between researchers from ICI and Haworth, Wiggins and Simonsen and an agreement was made in which ICI would provide Haworth with samples of catalysts for his work along with some confidential information, and the CPRC would supply ICI with some bulk samples of colonial products that were in short supply with the outbreak of war.[67] ICI suggested that the results of the work of the CPRC should be made available to all firms through the Association of British Chemical Manufacturers, as was the practice of the DSIR's Chemical Research Laboratory at Teddington. This arrangement was intended to remove any suggestion that the CPRC was providing unfair advantage to ICI.

The CPRC sponsored research into sugar derivatives at Birmingham University, under the direction of Haworth and his former student, Leslie Wiggins. Their team focused on the chemical reactions of sucrose and they generated large numbers of compounds that were then tested for any useful properties, particularly analgesic, chemotherapeutic or plasticising effects. The number of compounds generated by this approach could be substantial, with over 100 new substances produced from levulinic acid, a derivative of sucrose, in 1945/46.[68] By the following year, the researchers at Birmingham were being assisted in the laborious task of screening these compounds to find useful substances by the Department of Pharmacology at the University of Oxford and by a researcher at the Physiology Department, Birmingham.[69] In 1945 the research at Birmingham was extended to include the chemistry of starch, under the supervision of Wiggins and Stanley Peat. Starches from a number of colonial sources were investigated, including cassava and arrowroot from the Windward Islands, and a researcher was appointed in East Africa to survey sources of starch in this region.[70] In 1948 starch research was transferred from Birmingham to the University of North Wales at Bangor on the appointment of Peat to a position there.[71]

The CPRC also sought to identify new uses for natural commodities that had seen their markets undermined by the development of synthetic alternatives. One example was the flavouring vanillin. This substance had previously been derived as a natural product from clove oil exported from Zanzibar, but synthetic alternatives to vanillin were now being produced more cheaply from either guaiacol, a coal tar product, or from plant lignin. The possible collapse of the market for natural vanillin was considered especially worrying since the economy of Zanzibar was almost entirely reliant on the export of cloves. The CPRC decided to address this particular problem by initiating research into eugenol, the main component of clove oil, at King's College at Newcastle upon Tyne.[72] Similarly, the market for lime juice from the West Indies as a source of citric acid had been eroded through the development of fermentation processes for producing citric acid from molasses. Professor Ian Heilbron, who sat on the CPRC, was asked to embark on an investigation of the components and chemistry of lime oil, as well as other citrus oils, in his laboratory at Imperial College.

In the body of their annual reports, the CPRC laid much emphasis on the fundamental nature of the work they sponsored, saying, for example, 'It has been recognised from the outset that it would prove extremely difficult to find new uses for eugenol and its derivatives, and this can only result from fundamental research.'[73] The generation of large quantities of new compounds through fundamental research and the subsequent screening of these chemicals was a highly speculative approach to the discovery of useful substances. New compounds were tested for their suitability for a considerable range of new uses and this took enormous time and effort. The research sponsored by the CPRC was lengthy, laborious and unpredictable, and the council frequently cautioned readers of its annual reports against the expectation that rapid results might ensue from this work.[74] This particular characterisation of fundamental research as long-term and in-depth fundamental study can be seen as an argument against interfering in the work of scientific researchers by setting targets or trying to force early results. In the rhetoric of the annual reports of the CPRC, high-quality research was assured as scientists were afforded the opportunity to pursue their studies in the way they saw fit.

In the reports of the CPRC the term 'fundamental research' was also used to refer to work that explored the most general, basic, chemical reactions of a compound.[75]

> From their initiation, it was recognised that the experiments having as their object the finding of alternative uses for eugonol and vanillin were highly speculative and the most promising lines of attack lay in the study of the general chemical reactions of these substances.[76]

And

> In view of the great importance which the sugar cane crop has for the economy, not only in the West Indies, but also in other parts of the Colonial Empire, the council has decided that an investigation of the reactions of sucrose (cane sugar) should be started on a broad basis.[77]

The use of 'fundamental research' to denote work of a broad or generic nature was significant when it came to determining the appropriate role of government in the promotion of industrial development. The CPRC did not fund the study of limited and narrow lines of scientific enquiry as this type of study was likely to be of benefit to only a small number of firms. Using state funds in this way compromised the operation of market competition as it advanced the interests of one company over another. The exploration of general principles in the chemistry of materials such as sugar was acceptable, however, as the results were of potential benefit to an entire sector of chemical manufacturing. Individual companies would be able to take up the new information that came from investigations in sugar research and develop those findings along lines that were particular to the business of that firm. The focus on general investigations, then, provided a rationale for state-funded laboratory research of benefit to industry by allowing intervention by the state in the process of economic development while still leaving the initiative largely in the hands of business. This was a vision of the relationship between scientific research and economic development that was liberal in character.

Not all the work sponsored by the CPRC amounted to such long-term and basic elucidation of the chemistry of natural products. Other research projects overseen by the CPRC were concerned with making use of previously unexploited natural resources available in the Colonial Empire. The surveying and assessment of colonial products in this way was not dissimilar from the kind of investigations that had been carried out at the Imperial Institute as part of its commercial intelligence service.[78] At Liverpool University, research into the composition of a range of colonial fats and oils, including oils from linseed, citrus seeds, rubber seeds, wheat germ, groundnuts and sunflowers, was done under the direction of Professor T. P. Hilditch, in order to determine their commercial value.[79] In a similar vein, the CPRC funded the survey of plants from the colonies in the laboratory of Professor A. R. Todd at Cambridge University, in order to ascertain if they had any useful insecticidal or medicinal effects. Other projects included research into uses for colonial timbers and their resins at the DSIR's Forest Products Research Laboratory, Princes Risborough, and an investigation to find a use for theobromine, a by-product of the cocoa industry, at the University of Manchester. The CPRC also contributed to the post-war search for a British source of cortisone by evaluating the amount of ergosterol yielded by different strains of yeast.[80] Corticosteroids had been discovered in 1949 to be effective in the treatment of rheumatoid arthritis but the dollar shortage had forced Britain to find a way to manufacture its own cortisone rather than rely upon imports from the US.[81]

When it came to the research prosecuted in British universities, a relationship with industrial development was more assumed than overseen. The CPRC's Director of Research, Simonsen, made direct contact with Cadbury's, ICI, Unilever, Trinidad Leaseholds, Glaxo and Boots during the 1940s in order to publicise the results of the council's work. The CPRC, however, generally followed the model of the research councils in Britain when it came to the dissemination and uptake of its findings. The CPRC publicised its results and then it was up to business to make use of this information. The annual reports of the CPRC described the work undertaken with funds from the council with a list of papers in scientific journals and patents. During the 1940s, these were dominated by the sugar and starch research at Birmingham and Bangor.[82] At the end of the 1940s, the CPRC took further steps to ensure that its work contributed to the development of the British Caribbean by creating two new laboratories in Trinidad, close to producers of sugar.

The CPRC was created with the aim of contributing to the economic development of Britain's colonies after 1940. By finding new markets for tropical products, the council hoped that it would help to improve the poor social conditions that existed in places such as the British West Indies. The Secretary of State for the Colonies, Oliver Stanley, told the CPRC that 'the welfare of 60,000,000 people depended on the success of this work'.[83] These development objectives were to be achieved, in part, by state-funded research in British universities that worked out in detail the chemical constitution and reactions of products from the empire. Responsibility for allocating funds for the research programme lay with the CPRC in London, and research projects did not arise through requests by colonial governments or administrative staff of the Colonial Office, or in response to any enquiry from business. The researchers that received funds from the

CPRC were free to undertake time-consuming and long-term projects of work along lines they decided were appropriate. Some of the development needs of Britain's colonies, and aspects of their future economic growth, were determined by a group of eminent British chemists in London who were afforded a substantial degree of autonomy in making their decisions by the Colonial Office during the 1940s.

Conclusion

In 1940 a new approach to colonial development was launched at the Colonial Office that emphasised the necessity of centrally conceived and implemented projects. The 1940 CDW Act was given an important function in restoring Britain's reputation as an imperial power after the revelations of social deprivation and economic stagnation in the colonies that emerged by the late 1930s. A large commitment to scientific research was presented as evidence that in the future Britain sought to place development on a sure foundation of knowledge. As well as a practical necessity, the creation of a dedicated research fund was explained to the Treasury as being politically expedient. A major new commitment to comprehensive colonial research signalled to critics of previous colonial policy the willingness of Britain to take real action to remove obstacles to development in the colonies.

The idea that scientific research was the essential first step in a longer process of development appeared to have been belied in practice, however, by the distance that was created between the work of scientific researchers and the other functions of colonial governments or the operations of industry, as new apparatus for colonial research was introduced during the 1940s. If the concept of 'fundamental research' had symbolic value for officials, it had rhetorical power for the elite scientists who advised the Colonial Office. New research committees, including the CPRC, used the idea that the successful prosecution of fundamental research required particular working conditions to persuade officials to introduce arrangements for colonial research that aimed at freedom for research workers from oversight by other categories of officer. This was a discourse on the nature of research that served to legitimise and naturalise the preferences of elite scientists drawn from Britain's research councils. Colonial Office officials accepted claims of the necessity for freedom for researchers and central direction of research because of a belief that this was essential for the recruitment of high-quality scientists.

The creation of programmes of fundamental research into sugar and other tropical products was cast as having an important function in restoring prosperity to colonies that were overly dependent on the export of agricultural materials. Long-term and exploratory laboratory research aimed to identify new uses in chemical manufacturing for commodities in oversupply. The means by which knowledge would move from British universities and be taken up by business was not an issue that was well elaborated by the CPRC during the 1940s. As with the domestic research councils, the CPRC focused its efforts on creating the necessary conditions for research, therefore, it said, ensuring the high quality of this work, and it remained for business to capitalise upon the results of these investigations. By the end of the 1940s, however, the CPRC decided to create two new laboratories in Trinidad that undertook research into sugar and other products in order to help spur industrial development. The next chapters consider the relationship between the work of these laboratories and policy for the industrialisation of the British Caribbean as it evolved in the post-war period.

Notes

1. E. B. Worthington, *Science in Africa: A Review of Scientific Research Relating to Tropical and Southern Africa* (Oxford: Oxford University Press, 1938), p. 2.
2. TNA, CO 847/13/13.
3. Worboys, "Science and British Colonial Imperialism", pp. 221–230.
4. Worboys, "Science and British Colonial Imperialism", p. 226; J. Hodge, "Science, development and empire: the Colonial Advisory Council on Agriculture and Animal Health", *Journal of Imperial and Commonwealth History* 30 (2002), 5–9; Hodge, *Triumph of the Expert*; TNA, CO 927/12/3 gives some history of Amani.
5. C. Jeffries, *A Review of Colonial Research, 1940–1960* (London: HMSO, 1964), pp. 13–18.
6. Worthington, *Science in Africa*, pp. 320 and 341.
7. Frank Stockdale commented in his role as Agricultural Adviser at the Colonial Office in 1938 that the creation of a research fund would be the perfect opportunity to bring into effect the recommendations made by the Lovat Committee in 1927 for an expansion of research institutions across the Colonial Empire.
8. For a full account of the genesis of the African Research Survey, see H. Tilley, *Africa as a Living Laboratory: Empire, Development and the Problem of Scientific Knowledge, 1870–1950* (Chicago: The University of Chicago Press, 2011), ch. 2, and J. Cell, *Hailey, A Study in British Imperialism, 1872–1969* (Cambridge: Cambridge University Press, 1992).
9. R. D. Pearce, *The Turning Point in Africa: British Colonial Policy, 1938–1948* (London: Frank Cass, 1982), p. 43.
10. M. Hailey, *An African Survey* (Oxford: Oxford University Press, 1938), pp. xxix–xxv.
11. The allocation for research for the MRC for the years 1945 to 1950 was as follows: 1945–46, £295,000; 1946–47, £465,000; 1947–48, £698,000; 1948–49, £1,135,000; 1949–50, £1,618,000. The allocation for research for the ARC for the same period was: 1945–46, £300,000; 1946–47, £300,000; 1947–48, £400,000; 1948–49, £450,000; 1949–50, £777,000. The annual allocation for the Research Fund under the CDW Act of 1945 was £1,000,000.
12. The size of this fund makes its absence from accounts of the development of state-funded research in Britain during the twentieth century all the more surprising. The main accounts are P. Gummett, *Scientists in Whitehall* (Manchester: Manchester University Press, 1980); Wilkie, *British Science and Politics Since 1945*; S. Rose and H. Rose, *Science and Society* (London: Penguin, 1969); J. B. Poole and K. Andrews,

The Government of Science in Britain (London: Weidenfeld & Nicolson, 1972); N. J. Vig, *Science and Technology in British Politics* (Oxford: Pergamon Press, 1968).
13 Cell, *Hailey*, p. 237; TNA, CO 847/13/13.
14 TNA, CO 847/13/13.
15 *Ibid.*
16 G. B. Masefield, *A History of the Colonial Agricultural Service* (Oxford: Clarendon Press, 1972), p. 42; Worboys, "Science and British Colonial Imperialism".
17 Masefield, *History of the Colonial Agricultural Service*, p. 42.
18 TNA, CO 850/180/6.
19 D. Edgerton, *Warfare State: Britain, 1920–1970* (Cambridge: Cambridge University Press, 2005), ch. 3.
20 J. Austoker and L. Bryder (eds), *Historical Perspectives on the Role of the MRC: Essays in the History of the Medical Research Council of the United Kingdom and its Predecessor, the Medical Research Committee* (Oxford: Oxford University Press, 1989).
21 E. Mellanby, *The State and Medical Research* (Edinburgh: Oliver and Boyd, 1939), p. 15–19.
22 TNA, CO 900/1.
23 TNA, CO 874/36/4.
24 TNA, CO 927/1/3.
25 TNA, CO 927/1/3.
26 TNA, CO 847/13/13.
27 Edward Mellanby (1884–1955), research student under Frederick Gowland Hopkins, Cambridge University, 1905–07; qualified in medicine, St Thomas's Hospital, 1909; lecturer and professor in physiology, King's College for Women, 1913–20; Professor of Pharmacology, Sheffield, honorary physician, Royal Infirmary, 1920–33; secretary MRC, 1933–49; Fullerian professor, Royal Institution, 1936–37.
28 William Whiteman Carlton Topley (1886–1944), assistant director of pathological laboratory, St Thomas's Hospital, 1910; Director, Pathology Department, Charing Cross Hospital, 1911–22; Professor of Bacteriology, University of Manchester, 1922–27; Professor of Bacteriology and Immunology, University of London and Director, Division of Bacteriology and Immunology, LSHTM, 1927–41; member of MRC, 1938–41; secretary of ARC, 1941–44.
29 Edward Appleton (1892–1965), assistant demonstrator, experimental physics, Cavendish Laboratory, 1920; Wheatstone Professor of Physics, University of London, 1924–36; Jacksonian Professor of Natural Philosophy, Cambridge, 1936–39; secretary, DSIR, 1939–49; Nobel Prize for Physics, 1947; President of British Association, 1953; President, Radio Industry Council, 1955–57.
30 Archibald Vivian Hill (1886–1977), Fellow Trinity College and then Kings College, Cambridge, 1910–25; Brackenbury Professor of Physiology, University of Manchester, 1920–23; Nobel Prize, Physiology and Medicine, 1922; Jodrell Professor of Physiology, University College, London, 1923–25; secretary of the Royal Society, 1935–45; member, University Grants Committee, 1937–44; Chairman, executive committee, National Physical Laboratory, 1939–45; Scientific Adviser, Government of India, 1943–44; chairman, Research Defence Society, 1940–51; and a number of other committees during the 1950s.
31 Lee and Petter, *The Colonial Office, War and Development Policy*, p. 184.
32 On the involvement of Topley and Mellanby in the development of a bacteriological research programme, see B. Balmer, *Britain and Biological Warfare: Expert Advice and Science Policy, 1930–65* (Basingstoke: Palgrave, 2001).
33 In 1948 the Colonial Research Committee underwent a change of name to the Colonial Research Council with some changes in its membership. King was appointed to the Colonial Research Council at this time (representing the Advisory Council on Scientific Policy) until leaving on his appointment as Chief Scientific Officer at the DSIR in 1950.
34 A. King, *Science and Policy: The International Stimulus* (Oxford: Oxford University Press, 1984), p. 11, quoted in P. Chaston, "Gentlemanly Professionals within the

Civil Service: Scientists as Insiders During the Interwar Period" (DPhil, University of Kent at Canterbury, 1997).
35 TNA, CO 859/40/13; CO 859/79/13.
36 Jeffries, *Review of Colonial Research*.
37 *Colonial Research, 1942–1943*, Cmd 6486.
38 TNA, CO 852/588/2.
39 *Colonial Research, 1942–1943*, Cmd 6486.
40 *Colonial Research, 1942–1943*, Cmd 6486.
41 Edgerton, *Warfare State*, p. 348.
42 *Colonial Research, 1943–1944*, Cmd 6535.
43 TNA, CO 927/88/5.
44 TNA, CO 927/88/6.
45 TNA, CO 900/2; *Colonial Research, 1945–1946*, Col. No. 208.
46 *Colonial Research, 1944–1945*, Cmd 6663; *Colonial Research, 1945–1946*, Col. No. 208; *Colonial Research, 1946–1947*, Cmd 7151; *Colonial Research, 1947–1948*, Cmd 7493; *Colonial Research, 1948–1949*, Cmd 7739; *Colonial Research, 1949–1950*, Cmd 8063; *Colonial Research, 1950–1951*, Cmd 8303.
47 TNA, CO 927/88/6.
48 TNA, CO 847/36/4.
49 Lee and Petter, *The Colonial Office, War and Development Policy*, p. 171; Havinden and Meredith, *Colonialism and Development*, pp. 204–205; S. Clarke, "A technocratic imperial state? The Colonial Office and scientific research, 1940–1960", *Twentieth Century British History* 18(4) (2007), 453–480.
50 TNA, CO 852/588/2; Lee and Petter, *The Colonial Office, War and Development Policy*, pp. 171–172.
51 M. Worboys, "The Imperial Institute: the state and the development of the natural resources of the colonial empire, 1887–1923", in J. M. Mackenzie (ed.), *Imperialism and the Natural World* (Manchester: Manchester University Press, 1990), pp. 164–186.
52 G. Roberts and A. Simmonds, "British chemists abroad, 1887–1971: the dynamics of chemists' careers", *Annals of Science* 66 (2009), 103–128. The empire as a destination for chemists is a relatively neglected area of historical enquiry.
53 Havinden and Meredith, *Colonialism and Development*, p. 150. Worboys, "Science and British Colonial Imperialism", p. 250.
54 Worboys, "The Imperial Institute", pp. 164–186; S. Constantine, *Buy and Build: The Advertising Posters of the Empire Marketing Board* (London: HMSO, 1986), pp. 2–3.
55 P. Kumar, "Plantation science: improving natural indigo in Colonial India, 1860–1913", *British Journal for the History of Science* 40 (2007), 532–553.
56 C. Newbury, "Trade and technology in West Africa: the case of the Niger Company, 1900–1920",*The Journal of African History* 19 (1978), 551–575.
57 Worboys, "Science and British Colonial Imperialism", p. 355.
58 TNA, CO 899/1.
59 *The Manchester Guardian*, "Chemicals in War" (25 May 1945), p. 4.
60 S. Roskill, *Hankey: Man of Secrets* (London: HarperCollins, 1974), vol. III, pp. 363–601.
61 TNA, CO 852/506/11.
62 TNA, CO 899/1.
63 *Ibid.*
64 *Ibid.*
65 Reader, *Imperial Chemical Industries*, vol. II, p. 81.
66 P. Morris, *The American Synthetic Rubber Research Programme* (Philadelphia: University of Pennsylvania Press, 1989).
67 TNA, CO 899/1.
68 *Colonial Research, 1945–1946*, Col. No. 208.
69 *Colonial Research, 1946–1947*, Cmd 7151.
70 *Colonial Research, 1945–1946*, Col. No. 208.
71 *Colonial Research, 1948–1949*, Cmd 7739.

72 Colonial Products Research Council, *5th Annual Report, Colonial Research, 1947–1948*, Cmd 7493.
73 *Ibid.*
74 *Colonial Research, 1943–1944*, Cmd 6535; *Colonial Research, 1945–1946*, Col. No. 208.
75 See Chapter 5 of this book.
76 *Colonial Research, 1945–1946*, Col. No. 208.
77 *Colonial Research, 1946–1947*, Cmd 7151.
78 Worboys, "Science and British Colonial Imperialism", pp. 150–164.
79 S. Horrocks, "Industrial chemistry and its changing patrons at the University of Liverpool, 1926–1951", *Technology and Culture* 48 (2007), 43–66.
80 *Colonial Research, 1944–1945*, Cmd 6663; D. Cantor, "Cortisone and the politics of empire: imperialism and British medicine, 1918–1955", *Bulletin of the History of Medicine* 67 (1993), 463–493.
81 V. Quirke, "Making *British* cortisone: Glaxo and the development of corticosteroids in Britain in the 1950s-1960s", *Studies in the History and Philosophy of Biological and Biomedical Sciences* 36 (2005), 645–674.
82 *Colonial Research, 1947–1948*, Cmd 7493.
83 TNA, CO 899/1.

CHAPTER THREE

'Men, money and advice' for Caribbean development

In a break with previous policy, the Colonial Office announced in 1943 that it would promote industrial development in Britain's colonies. Manufacturing ventures were now deemed essential to raise living standards and address the politically dangerous issue of colonial unemployment. Officials became occupied with the question of what constituted acceptable modes of intervention by metropolitan and colonial governments to facilitate economic diversification. The challenge was to reconcile the need for demonstration of a more constructive approach to development with some long-standing laissez-faire principles. Two wider political issues made Colonial Office attempts to persuade the Caribbean colonies to follow its preferred routes to industrialisation difficult, however. The increasing political autonomy of governments in the Caribbean region meant that Britain could not merely instruct its West Indian possessions to follow its edicts. In addition, it became clear that in the post-war world, the US hoped to shape development across the Caribbean along lines that it found conducive to its own interests. In the face of these challenges, the collation and dissemination of economic advice assumed an important role in the maintenance of Britain's control over its British West Indian colonies after 1940.

After 1944 Britain was involved in direct negotiations with the US about region-wide policies for the West Indies as a member of the Caribbean Commission. Despite the ostensible collaboration represented by this body, discussions between officials reveal substantial and entrenched differences in British and American visions for the industrial development of the Caribbean. These differences proved difficult to reconcile, informed as they were by the wider economic and political priorities of Britain and America in the post-war world. From the perspective of the British government, the actions of the Caribbean

Commission, the creation of Truman's Point Four Programme and the UN system held out the prospect that in the post-war period the process of decolonisation would be one in which power and influence in the region moved out of British, Dutch and French hands into the hands of the US through the activities of its advisors and businessmen.

The fact that more than one path towards industrial development was promoted to the territories of the Caribbean during the 1940s and early 1950s is important in helping us rethink the picture of development we have for the period after 1945. No single coherent post-war development paradigm shaped the thinking of experts and officials from Britain and the US when it came to ways to encourage manufacturing in the region. The visions that were promoted pre-dated the advent of modernisation theory and W. Arthur Lewis's landmark work on economic development.[1] American and British officials drew upon a range of historical events and contemporary experiences, consulted economists and businessmen, and carried out surveys, in order to formulate and legitimise their preferences. The 1940s and early 1950s were a time when ideas about industrial development were being worked out and the Caribbean was a laboratory for the emergence of proposals for the stimulation of economic diversification in nations that had previously been dominated by the production of primary goods.

Plans for industrial development in Britain's colonies

Expert advisors assumed a great deal of importance in the formulation of British colonial policy after 1940. The Colonial Office had seen the appointment of increasing numbers of specialist officers and committees in the interwar years, but their recommendations did not necessarily amount to significant metropolitan interference in colonial affairs, not least as advice was not matched by large amounts of funding for initiatives. With the passing of the 1940 CDW Act, officials set about creating new bodies in London that would assume a greater role in the production of development plans for Britain's colonies. Sydney Caine, Assistant Under-Secretary and Head of the Economics Division in April 1944, advocated a more active approach by London to getting projects off the ground so as to halt the drift in policies of colonial governments that had occurred in the interwar period. Experts in scientific research and economics at the Colonial Office were charged with creating new and effective interventions in the field of development.[2]

The Colonial Office began discussion about colonial industry with the turn to consideration of post-war reconstruction in 1943. Debate about industrialisation was inaugurated by a speech in the House of

Commons in July 1943 by the Secretary of State for the Colonies, Oliver Stanley, in which he declared that the colonies would never reach a reasonable standard of development without some degree of industrialisation. Stanley urged caution, however, with regard to the scale of this change and the methods to manage industrial development. The use of tariffs against imported goods in order to protect fledgling industry was singled out for special condemnation. Stanley stated in his speech, 'I cannot think of anything more fatal to the economics of the Colonies than a rash, mushroom, industrialist growth, fostered by high protective tariffs unrelated to either local products or local markets.'[3]

L. J. Butler says that in the period before 1940, the Colonial Office was not opposed to industrialisation per se but had firm objections to tariffs on the basis that they raised prices for colonial consumers, reduced colonial government revenue from imports and could adversely affect British manufacturers.[4] In 1934 an interdepartmental committee considering colonial industrialisation had rejected government intervention to stimulate industrial activity, stating, 'there is no reason at all why a conscious policy of artificial encouragement of industry should be undertaken and pursued'.[5] While continuing to reject certain 'artificial' measures to encourage industry, the speech by Stanley in 1943 marked a shift in policy. Stanley's call for a more assertive approach to developing industry was prompted by the crisis in the British Empire in the 1930s. The Great Depression had shown that too many of Britain's colonies were dependent on a narrow range of agricultural exports, making them highly vulnerable to the fluctuations of the world market.[6] The encouragement of colonial industry was a way to solve the issues of unemployment and low living standards. In further contrast to the recommendations of the interwar period, Stanley claimed that the new policy for industrialisation would not prioritise the interests of British manufacturers. New colonial industry could focus on producing simple cheap goods such as soap and textiles whilst British firms would be engaged in more skilled forms of manufacturing. The development of colonial industry producing simple products would benefit British manufacturers as higher-value goods from Britain would find a market in the colonies amongst colonial consumers who had acquired greater purchasing power as their own economies developed. Stanley's final point in his speech concerned the respective roles of government and private capital in the establishment of industry. He told Parliament that the new 1940 CDW Act would address the need to provide a suitable background for industrialisation by funding the development of infrastructure, services and social improvement.[7]

Stanley left open the question of how exactly the Colonial Office in London and colonial governments were to encourage the industrial development he envisaged. The issue was put before the Colonial Economic Advisory Committee (CEAC) that had been created in September 1943 to provide expert advice on economic aspects of development. CEAC included amongst its high-powered members the eminent economists Hubert Henderson and Lionel Robbins, who were economic advisor to the Treasury and Director of the Economic Section of the War Cabinet, respectively.[8] Two other economists, both from the London School of Economics (LSE), were also appointed: the socialist and Labour Party member Evan Durbin and the St Lucian scholar W. Arthur Lewis, who took on the role of Secretary. Lewis had joined the faculty of LSE in 1938 while finishing his PhD and combined his teaching responsibilities with a number of ad hoc roles at the Colonial Office during the 1940s, for which he produced a large number of reports and memoranda that often prefigured his later work in development economics. During the lifetime of CEAC, Robbins and Henderson did not attend many meetings and debate was often driven by the contributions of Durbin and Lewis. The early meetings of CEAC were dominated by the question of colonial industrialisation, with clear points of contention emerging early on, particularly when it came to the views of Durbin and Sydney Caine.

Caine has been identified by historians as a figurehead for the new way of doing things at the Colonial Office that emerged with the passing of the 1940 CDW Act.[9] Caine's belief was that poor economic performance and inadequate levels of social provision in the colonies could be addressed by the introduction of a degree of planning, but the absence of men with specialist skills in the colonies, including qualified economists, presented an obstacle. The solution was to place the responsibility for the creation of plans in the hands of metropolitan bodies that had the necessary expertise. In a well-known memorandum of 1943, Caine singled out the CRC as an example of a body that took the initiative and formulated projects of colonial research itself: 'the Committee has with great emphasis repudiated the idea of confining itself to the merely negative or censorial function of passing judgement on schemes devised by other people and submitted to it'.[10] As with scientific research, the appointment of a metropolitan committee of eminent economists in the form of CEAC was an attempt to provide direction in another area where colonial governments were said to have few skills and little experience. In the words of CEAC,

> The Committee can say only that, if the pace of economic development is to be adequate, some means must be found of imparting to local or

regional planning authorities from the centre those larger directives which they are unaccustomed to formulating of themselves.[11]

Despite his desire to see a faster pace of change in the colonies and his enthusiasm for greater interventions by well-qualified bodies in London, Caine was not an advocate of a high degree of government control over the process of industrial development. Indeed, although a great deal of talk about planning can be discerned in the files of the Colonial Office around 1943 when officials took up the question of post-war reconstruction, the sort of planning that was envisaged was not explicitly articulated. The debate about industrialisation in the colonies reveals much about the limited definition of Colonial Office planning in practice. While Caine urged his colleagues to take a more active approach to formulating and implementing policy for the colonies than the Colonial Office had traditionally done, he was arguing for a rejection of absolute laissez-faire approaches. He was not advocating that in future, government would control industries or oversee the allocation of resources and manpower. Planning for Caine was only ever 'outline' planning.[12] In contrast, Durbin and Lewis were keen to see the Colonial Office and colonial governments assume a much bigger role to intervene and force rapid industrialisation in selected areas of the Colonial Empire. Since this was a proposal at odds with the course favoured by Caine, CEAC struggled to reach consensus and formulate a report that could be considered as the basis of advice from the Secretary of State to the Colonies. Meanwhile, a conference was held in the Caribbean at which the issue of industrialisation figured prominently.

Britain and the Anglo-American Caribbean Commission, 1942–45

Between 21 and 30 March 1944 the first West Indian Conference of the Anglo-American Caribbean Commission was held in Barbados. The Caribbean Commission had been established in March 1942 as a joint enterprise between the US and Britain, but its origin lay in proposals by the US government. In its first incarnation the Caribbean Commission was co-chaired by Frank Stockdale, who was Comptroller of Development and Welfare in the British West Indies, and Charles W. Taussig of the US State Department. Taussig was close to Roosevelt and had been one of the original members of the President's 'brains trust', a circle of advisors that worked on the New Deal. He was a businessman as well as a political advisor and had first-hand experience of the Caribbean as President of the American Molasses Company.[13]

In 1941 the US signed a 99-year lease with Britain that allowed US military bases to be established in six British colonies, including Trinidad. The stationing of American soldiers in the British Caribbean was only a part of the US strategy for the security of its 'backyard'. The US also wished to see programmes of social, economic and political reform introduced in the Caribbean in order to alleviate the poverty that was considered to fuel social unrest and spread communism. The US had two possessions in the region: Puerto Rico and the Virgin Islands. Puerto Rico, acquired by America in 1898, had achieved notoriety by the interwar period for its deplorable living standards. The slums, unemployment and overdependence on the sugar industry seemed intractable problems before the war despite a number of initiatives by American governors of the island.[14] In the early 1940s, a further extension of the New Deal to Puerto Rico was deemed necessary demonstration of enlightened American attitudes towards colonial governance. Beyond Puerto Rico, American-led improvements in the social and economic conditions of the colonial peoples of the entire Caribbean region were considered necessary to prevent uprising amongst the poor and unemployed. There was particular concern about the consequences for regional stability of continued unrest in Britain's larger colonies. Historians have noted the role played by the Caribbean Commission in hastening political reform in Jamaica, where a new constitution that provided universal suffrage was introduced in 1944 after pressure on Britain from Taussig and other US officials. The US government declared its commitment to ensuring that self-government was brought to all dependent peoples.[15] Apart from the need to avoid riots in an area of importance during the war, US officials were concerned by the contacts that had developed between the black populations of the Caribbean area and African Americans at home. The emergence of the Harlem nexus meant that protest in one place could fuel discontent in the other.[16] Officials therefore linked a pressing need for economic, political and social reform in the Caribbean area to the domestic security of the US.[17]

In 1941 Taussig had been given permission by Britain to undertake a general survey of the British West Indies, and the resulting report was sent to the British government. In return the US received a copy of the unpublished Moyne Report, and this resulted in the suggestion by Taussig that since many similarities existed between the problems of Puerto Rico and the British West Indian colonies, Britain and the US should form a joint commission to formulate ideas that could solve common problems.[18] The Anglo-American Caribbean Commission was subsequently created with the expressed aim that it was a body formed to undertake studies and make recommendations that addressed the problems of 'labor, agriculture, housing, health, education, social welfare, finance, economics and related subjects'.[19] In 1946 the name of the Commission was changed to the Caribbean Commission as France and the Netherlands joined the body. The Caribbean Commission created four sections for its work, including the Caribbean Research Council formed in 1943 to collate data of use to governments of the region and which published the *Caribbean Economic Review*. The future president of Trinidad, Eric Williams, was appointed secretary to this body in 1943 as a member of the British Section.

The agenda of the first West Indian Conference convened by the Caribbean Commission in 1944 included industrial development. Delegates debated the issue of whether or not governments should provide capital for industry if private sources were not forthcoming; whether government should finance research; and whether tariffs, import duty relief and income tax relief were appropriate techniques to encourage new industry.[20] The US Section submitted papers that promoted the programme for industrial development that had recently begun in Puerto Rico. These included a copy of a speech by Rexford Tugwell, the Governor of Puerto Rico, given in February 1942. Tugwell was an academic economist, a friend of Taussig and another original member of Roosevelt's 'brains trust' whose outspoken commitment to the New Deal had earned him the name 'Red Rex' in Washington.[21] Tugwell's speech celebrated the achievements of the public Puerto Rico Development Company, usually referred to as either PRIDCO or Fomento. From 1942, PRIDCO had received an allocation of half a million US dollars each year from the Puerto Rico government for the establishment of new industry,[22] and a subsequent act created the Puerto Rico Development Bank that used public funds to issue loans to business.[23] The finance for these initiatives came from a rebate on rum import taxes paid to the colony by the US government that proved to be a substantial sum during the war when consumption of rum increased significantly as other spirits were in short supply.

The new development agency and bank in Puerto Rico and the programme for industrialisation that they administered and funded were the creation of Teodoro Moscoso. Moscoso was born in Spain to a Puerto Rican father and had been educated in New York and Puerto Rico before studying at the University of Michigan. On return to Puerto Rico in 1932, Moscoso had become involved in running a housing programme that cleared slums and built new houses for poor residents in the city of Ponce. A meeting with Luis Munoz Marin, who was elected President of the Senate in 1940 as leader of the Popular Democratic Party, was a major turning point for Moscoso, prompting him to join Munoz's party. In 1941 he took up a position as an aide to

Tugwell, and in the position of coordinator for insular affairs Moscoso created and directed an industrialisation programme for Puerto Rico. According to his biographer, Moscoso's ideas emerged after study of a number of foreign industrialisation programmes and discussions with individuals who had been engaged in formulating economic plans for Puerto Rico in the 1920s and 1930s. In 1942 Moscoso consulted the US chemical and engineering firm Arthur D. Little about his proposals and with the endorsement of the firm created PRIDCO. PRIDCO began by taking over a cement works and running it as a government factory before creating a glass works and a cardboard box plant, both industries related to the rum industry.[24]

The creation of glass and cardboard box factories meant overcoming the refusal of the War Production Board (WPB) in Washington to give a permit that would allow the importation of machinery to Puerto Rico. The situation was resolved after the intervention of the unlikely figure of Senator Robert Taft, known for his hostility to the New Deal and to Tugwell in particular. Taft's comments in a letter he wrote for the WPB are revealing for what they tell us about the exception that was being made for Puerto Rico when it came to the implementation of a state-controlled industrialisation programme, 'I have never been very strong for government-supported industry, but the situation in Puerto Rico is such that I believe the government had a proper function in promoting the development of new industry.' The well-publicised deprivation that existed in Puerto Rico was such a threat to US prestige that what was 'unacceptable socialism' for America was tolerable for Puerto Rico.[25] In April 1943 the WPB approved the construction of the glass plant and work began the same year.[26] During its early years the plant was beset with problems – machinery was stolen, the plant was badly designed, too many bottles were rejected as substandard and there were labour disputes. Similarly, the cardboard plant opened late as machinery was defective and businesses that had agreed to supply waste paper to the plant boycotted it.[27] Problems also arose with the next two government factories for shoes and ceramics.

Despite the bad publicity that was generated by the first initiatives of PRIDCO, details of the Puerto Rican programme were set out for the delegates of the West Indian Conference of 1944 and the Caribbean Commission promoted 'Operation Bootstrap', as it became known, as a model for Caribbean industrialisation throughout the 1940s and 1950s. The US had previously promoted its colony as a model of political reform, announcing in 1942 that plans were under way to allow the role of Governor to be an elected seat.[28] Political reform and radical economic policies to improve the lives of Puerto Ricans were essential if the US was to refute claims that in reality it was just another colonial power with no moral authority to pressurise Britain, for example, to commit to greater progress towards giving its colonial subjects the right to self-determination.

In response to the information on PRIDCO circulated in 1944 there were discussions at the Colonial Office on the question of whether or not similar development corporations were needed to provide expertise and maybe even capital for new industry for the British colonies.[29] In the summer of 1944, Simonsen and Robinson of the CPRC visited Puerto Rico as part of a trip to the Caribbean and saw for themselves some of the initiatives undertaken as part of the industrialisation programme. The subsequent discussions made a contribution to the decision to create the Colonial Development Corporation in 1948, to provide capital and participate in the running of agricultural and industrial projects, although by this point the priority was less the need for economic diversification in the Colonial Empire and more the urgent necessity of resolving Britain's economic crisis by earning dollars though the increased production of primary products.[30] Overall, the initiatives undertaken by Puerto Rico in the first half of the 1940s in terms of government development corporations, development banks and government-run factories were not endorsed by the Colonial Office as a model for industrial development for the British Colonies of the Caribbean. There was nothing the Colonial Office could do, however, to stop the Puerto Rican example continuing to be promoted within the Caribbean.

In the immediate wake of the first West Indian Conference of 1944, British officials were most preoccupied with the recommendation made by the Industrial Research Section of the Caribbean Research Council that this body should receive all plans for industrial development formulated in the region. This suggestion by the US Section was an attempt to create a central committee that coordinated industrial policy across both the British and American territories of the Caribbean.[31] Whilst exemplifying the core function of the Caribbean Commission as the US saw it, namely that it was a body to produce Caribbean-wide policy, this recommendation was the first of many to cause alarm at the Colonial Office. Britain had agreed to the formation of a joint commission for the Caribbean as it provided an opportunity to get British West Indian issues a higher profile in Washington and address some pressing wartime problems such as the shortages of food that were causing distress in the British colonies.[32] In private, officials in London were dismissive of the value of the work of the Caribbean Commission and its Caribbean Research Council, referring to 'the so-called Industrial Research Sub-Committee of the Caribbean Research Council' and commenting that 'none of the local peoples had two ideas to rub together', seemingly a reference to representatives from the Caribbean colonies such as Moscoso, who had been appointed to the Caribbean Research Council on the invitation of the US.[33] The Caribbean Commission generated anxiety amongst British officials and experts, however, even if the quality of its ideas were derided, as the US Section appeared intent on expanding the powers of the body and thereby extending the influence of the US in the region. The Colonial Office was keen to ensure that the role of the Commission was confined to providing information, not policy. The suggestion at the conference that plans formulated in British territories should be routinely sent to the Industrial Research Sub-Committee of the Commission was firmly rejected by the Colonial Office for fear this would sideline the Colonial Development and Welfare Organisation based in Barbados (CDW Org) that had been created with the passing of the 1940 CDW Act. The role of the CDW Org was to provide expert advice to governments of the British West Indies as they formulated their plans for development.[34] Oliver Stanley directed that copies of all industrial plans were to be sent to Frank Stockdale, who was Comptroller for Colonial Development and Welfare at the CDW Org, and Stockdale would decide what information should then be communicated to the Caribbean Commission, ensuring that the authority of the British government was maintained over that of the new Anglo-American Commission.[35]

CEAC and the development of manufacturing industries

In the months after the West Indian Conference, Stockdale found himself in the embarrassing position of having no Colonial Office guidance on industrial development to offer the governments of the British Caribbean. In a circular for the British West Indian Territories of May 1944 he explained that the matter was before CEAC at the Colonial Office and he was awaiting their report.[36] Stockdale wrote a letter to Oliver Stanley in which he made it clear that a statement from the Secretary of State on general policy with regard to industrial development for the British Caribbean colonies was needed urgently.

> As you are aware, there is an insistent demand for industrial development in the Caribbean at the present time. It is of the greatest importance that such development should be properly planned and controlled and I hope that you will be able to make an early statement on the general policy to be followed in order that the West Indian Governments may proceed without delay to take such action as may be considered necessary.[37]

In London, CEAC was engaged in heated discussion about the development of secondary industry, prompted by the circulation of a memorandum written by Lewis.[38] Lewis had made a number of recommendations, including the need for agricultural development to occur alongside industrial development in order to release labour for work in secondary industry and the need for industry to develop on a regional basis. Most controversially, from the perspective of the Colonial Office, Lewis rejected the idea of the gradual evolution of industry, recommending instead a 'sudden jump'.[39] This suggestion prompted fierce debate at meetings of the committee, with Lewis and Durbin on one side, and Caine, Sir Bernard Bourdillon (former Governor of Nigeria and a director of Barclays Bank) and the Chair of CEAC Sir Harold Howitt (of the accountancy firm Peat Marwick, Mitchell and Company) on the other. The issues that divided the committee were the pace of change that was desirable, the idea of centres of mutually supporting industrial development and the degree of government intervention that was necessary. Lewis and Durbin supported a thoroughgoing programme of rapid change for Britain's colonies which the Colonial Office deemed unacceptable.

Of the economists on CEAC, Evan Durbin was the most vociferous supporter of the idea of rapid industrialisation. Durbin recommended that efforts should be concentrated on specially chosen areas of the British Colonial Empire, such as Nigeria, with the aim, 'to modernize it swiftly by deploying large economic resources for the region – after the TVA model or the alleged achievements of the USSR in its "colonial territories" '.[40] This proposal was firmly rejected by Caine, Bourdillon and Howitt, and instead Caine called for the rejection of the, ' "revolutionary" industrialization of selected territories', in favour of 'evolutionary development of a variety of industries in a large number of dependencies'. The idea of the gradual evolution of industry indicated that much of the initiative for the emergence of new manufacturing would lie with business rather than government. This was a vision of industrialisation in which, on the whole, manufacturing enterprises would find a footing only where market conditions allowed. The main role of governments was to provide public services and technical education and fund useful scientific research. Caine did concede, however, that it might be wise to provide special concessions to attract new industries in the short term, and in some cases he raised the possibility of provision of government capital to help establish a factory where private capital was absent.

Durbin's response was angry and he accused the Colonial Office of sticking to its traditional policy of inactivity when it came to the development of colonial manufacturing. The Colonial Office was effectively saying, according to Durbin:

> We believe that nothing can be done except to improve and expedite (without giving any estimate of the practicable increase of speed) our traditional policy. It is true that these policies have produced very little industrial development in most of the territories for which we are responsible during a period in which immense changes have taken place in other parts of the world, but we nevertheless think that this is the best that we can do.[41]

Durbin and Lewis continued to press for government intervention to spur rapid economic development, arguing that this should be the main aim of development policy, not least as it would ease the financial burden on the British government if the colonies became self-supporting in terms of social services. They claimed that a slow pace of a change for the colonies placed the Colonial Office out of step with public opinion in Britain. Most importantly of all, Durbin and Lewis accused Britain of failing to recognise the aspirations of colonial peoples since the so-called evolutionary approach to economic change favoured by officials would only retard the moment when self-government would be attained.[42] In an attempt to promote their cause, the two economists made a direct appeal in October 1944 to Oliver Stanley, asking for his endorsement of the concentration of rapid industrialisation in a few key areas, only to have this rejected.[43]

Lewis's last attempt to make a case to CEAC for the development of centres of industrialisation in the empire was a memorandum prepared with F. V. Meyer. 'The Analysis of Secondary Industries' stated that focused points of industrial development were the most efficient way to spend development money and most likely to provide an environment in which new factories might flourish. This document was notable for attacking a basic Colonial Office principle when it came to the question of industrial development. Meyer and Lewis stated that judgements at the Colonial Office on the desirability of establishing an industry were usually predicated on one question: will it be profitable? Whilst judging a new industrial concern on its individual merit might be reasonable in the case of the UK and US, Lewis and Meyer made the point that this could not be done in the colonies. In the colonial case, the cultivation of a whole industrial sector needed to be considered, as the profitability of one firm could change a great deal with a changing wider industrial context. They recommended that government focus on encouraging the setting up of trading estates where a group of industries could be encouraged to their mutual advantage. Lewis and Meyer's report therefore provided another iteration of the idea that industrial development needed to be actively encouraged in selected places.[44]

Lewis resigned from CEAC in November 1944, criticising Caine for rejecting the need for some government action to facilitate economic development. According to Lewis, Caine had given the examples of 'Britain and the USA as countries which developed rapidly without government prodding'.[45] Lewis described himself as astonished by this assertion since the situation in the colonies, which had little in the way of capital and entrepreneurs, was not comparable with Britain in the eighteenth century. For Lewis, industrial development would not naturally or spontaneously occur in the colonies; the economic development of these places required government planning. Caine in turn claimed that the breakdown in relations between CEAC and the Colonial Office was the fault of members who had overstepped the limits of the committee's original remit and considered political issues rather than restricting themselves to economic ones. In fact, it seems the problem was that Durbin and Lewis advocated policies that amounted to a direct attack on the laissez-faire principles that still underpinned the thinking of the Economics Department despite a new rhetoric of centralisation and planned development. In 1946 the committee was disbanded and replaced with a Colonial Economic and Development Council.[46]

The anticipated CEAC memorandum on industrial development was finally circulated to the colonies in February 1945. Despite the urging of Lewis and Durbin that the Colonial Office seize the opportunity for planned, intensive and rapid industrialisation of key areas of the Colonial Empire, this final document reflected the views of Caine. In his introductory letter to the circular Stanley stated 'No attempt has been made to suggest any very rapid or revolutionary changes', instead the memorandum outlined the 'steps which are open to Colonial Governments without such revolutionary change and which are likely to be acceptable as matters of general policy'.[47] The memorandum itself began with the statement that industrialisation was to be encouraged but that it would be wrong to offer state assistance that encouraged the 'artificial' development of industries that could not be economic in the long term. Industries that were permanently dependent on tariff protection or import restriction would not benefit the colonies; new industry should only be encouraged in the expectation that one day it could successfully compete with those of the more developed countries. Caine's memorandum gave a role to government in the removal of obstacles that might hinder industrial growth. The barriers to industrialisation that government needed to help surmount included a shortage of skilled workers and managers, the absence of basic services, or the provision of information through research. Government might also provide inducements to business such as duty-free imports of machinery. If there was clear expectation that a business would be profitable in the longer term then subsidies and protection in the short term could be acceptable. The report reminded the reader, however, that the most common form of protection, tariffs, had the effect of raising prices for the colonial consumer. The cost of using tariffs to help new industry ultimately fell upon 'those least able to bear it and whose standard of living it is the ostensible object of policy to improve'. The report left much to the discretion of colonial governments in determining the exact approach they would follow.[48] In October 1945 Stockdale circulated to the colonies of the British West Indies a short summary of Caine's recommendations in which he outlined five 'legitimate' forms of government help that might be extended to new industries: assistance with the costs of scientific research; duty-free entry of machinery; relief from taxation; training; and the improvement of transport facilities.[49] In general, the vision of industrialisation promoted by the Colonial Office was one in which development would occur gradually across the Colonial Empire through the actions of businessmen, with government helping to produce convivial conditions for private investment.

Further insight into the criteria applied by the Colonial Office when considering if an industry might be judged to be desirable is provided by the response to an enquiry from the Governor of Trinidad in January 1946. Bede Clifford requested advice on the incentives Trinidad might offer the British firm Associated Portland Cement Manufacturers (APCM) to set up a factory on the island.[50] The Colonial Office advised the Trinidadian government to say that it was prepared to allow the duty-free entry of machinery and equipment required to set up the plant in Trinidad and would consider providing transport facilities to the site of the works if appropriate.[51] Privately, officials posed the question, 'was it better for a Colony to buy its cement in the cheapest market because cement was so important to it, or to establish a local industry with all the benefits which that entails, but at the cost of a rather higher price for cement?'[52] Caine was asked to give his view on the matter. He described the situation as an 'awkward conundrum', and stated that there existed little consensus amongst economists on the issue of cost to consumers versus the value of the creation of new industry. He suggested to his colleagues at the Colonial Office that they might advocate a compromise.

> What might be described as the middle line of modern thought, lying between the extreme free trade view that protection is never economically

worth while and the extreme protectionist view that it is worth putting almost any burden upon the consumer in order to increase internal production, is that some protection to a local industry may be on balance advantageous if it results in employing local labour which would otherwise be without employment.

If a local cement industry meant an increase in prices or loss of revenue, it was not worth pursuing since the plant would require a large investment in machinery and would provide little new employment. Caine's conclusion was that the cement works did not provide much economic advantage to Trinidad and so the government should not attempt to persuade APCM to establish a factory there.[53] The wider point demonstrated by Caine's advice was that in the Colonial Office conception of economic development, industrialisation was not desirable merely for its own sake. Lewis and Durbin, on the other hand, spoke of industrialisation as a necessary step on the path to economic maturity and political freedom.

US and British visions of industrialisation, 1945–52

In the first round of discussions between 1943 and 1946, Britain had formulated principles for the encouragement of industry that it hoped its colonies would follow. This advice from the Colonial Office was not substantially revised during the 1940s. Indeed, in 1950 the Colonial Office stated that Caine's 1945 memo still stood as the most comprehensive expression of British policy on colonial industrial development and one that the Colonial Office said 'we still regard as adequate'.[54] The approach favoured by Britain, in which government interventions were cautious and limited, was challenged, however, by other members of the Caribbean Commission, who persisted in their efforts to coordinate and plan industrial development across the region. American attempts to see greater interventions to foster industry within the borders of individual colonies was coupled with a drive for a more liberal trade region between territories. The US Section pushed for trade agreements to facilitate the exchange of foodstuffs and manufactured goods in line with wider US policy to reduce tariffs across the world economy. There were also calls for government capital to be provided for a regional bank and for the Puerto Rican industrialisation model to be taken up by other territories. These latter suggestions found favour with officials in the French and Dutch Sections, but not at the Colonial Office. British officials were aware that the US intended to use the Caribbean Commission to extend its influence in the region and shape both the economic and political future of the possessions of the European empires along lines that confirmed to the objectives of

US foreign policy. On more than one occasion, US officials attempted to formally align the Caribbean Commission with the UN in some way, as a specialised agency or a regional council, in order that the territories of the region would abide by principles expounded by the UN such as the Universal Declaration of Human Rights. This ambition was frustrated by officials from France and Britain, who rejected the suggestion of a closer relationship with the UN for potentially undermining their authority over their colonies and as an attempt by the US to force them to a timetable for granting independence.

British officials only occasionally resorted to outspoken dissent in their meetings with other sections of the Caribbean Commission. They were often able to avoid making a commitment to new initiatives by prolonging the process of fact gathering, report writing and debate, the slow pace of which was exacerbated by the internal disputes that sometimes broke out between the other members of the Commission. In rejecting the suggestions of the Caribbean Commission, officials were asserting the primacy of British advice and British capital, both private and public, in the development of British territories. The idea of an integrated and organised Caribbean economy that transcended the borders of the colonial empires in the region was a key element of US policy as promoted by the Caribbean Commission through the 1940s and 1950s. This was in line with the wider ambitions of the US government for a liberalisation of trade in the post-war world and also to ensure that key raw materials produced in the Caribbean, such as bauxite, were made available to US business. It was also considered essential for US security; a more prosperous and unified Caribbean was considered necessary to withstand the threat of communism to an area described as the 'soft under belly of the United States'.[55] The fact that the US Section of the Commission promoted an entity, the wider Caribbean, was a problem for the Colonial Office. US attempts to forge the Caribbean as an economic unit were not easily reconciled with Britain's preferred political and economic categories – 'empire', 'the British West Indies federation' and 'commonwealth'.

The view of Franklin D. Roosevelt was that the Caribbean should be an economic unit, not of free trade, but of multilateral trade agreements, and he argued for wider economic planning so that different Caribbean locations did not act in competition with each other.[56] Both the Atlantic Charter of 1941 and the Mutual Aid Agreement of 1942 signed by Britain and the US spoke of the removal of discriminatory trade practices so as to avoid a repetition of a cycle of increasingly protectionist measures as had damaged the world economy during the Depression. The US Section of the Caribbean Commission believed that it needed to combat a tendency towards narrow nationalism in the

Caribbean and spoke of encouraging the colonies to act in accord on the world stage so as to strengthen their position with respect to more powerful economies. Taussig told Stockdale in 1942 how he imagined greater integration would work in practice:

> the economies of Puerto Rico and the Dominican Republic were in many respects complementary; that they can mutually benefit from the labor and purchasing power of the one and the surplus food production of the other; that similarly Jamaica, should it develop irrigation projects for the production of rice, may find a market in Cuba and Cuban beef in turn a market in Jamaica.[57]

While Britain agreed in a joint declaration of 1944 that 'The economic problems of the Caribbean should be regarded as regional rather than local problems', it transpired that Britain did not see this as a commitment to reducing imperial preference or the rationalisation of industry across the region.[58] Britain continued to maintain a system of preferences in the post-war period and then introduced additional controls on trade and currencies as Britain entered a period of economic crisis after 1947. This worked to frustrate the ambitions of the US Section to see the British colonies opened up to trade outside of the imperial system. When the Colonial Office considered reforms in tariff arrangements or the need to consider the development of industry beyond individual colonies, it did this with the projected federation of the British West Indies in mind. A Federation of the British West Indies, in which all Britain's Caribbean colonies would be brought together to form an entity that coordinated economic policy and created a customs union was one key way in which Britain aimed to protect its territories from the threat of American interference. The idea of a federation was first advanced in 1947 and debate was under way on such issues as the headquarters of the federation and the nature of its administration whilst the Colonial Office was engaged in debate with the US Section of the Caribbean Commission.

Records of conversations that occurred with Washington make it clear that US officials hoped that one of the main beneficiaries of trade across the Caribbean would in fact be Puerto Rico. Moscoso was seeking local markets for goods manufactured in Puerto Rico as part of his programme of industrial development and sought a reduction of shipping costs to facilitate this: 'it is imperative that facilities for freight transportation be established. If we are ever to become the workshop or industrial center of the Caribbean *we must* be able to ship some of our surplus products to other Caribbean islands.'[59] He also raised the issue of the refusal of the Trinidad Control Board to grant a licence to import cement from Puerto Rico. The British Section of the Caribbean Commission informed the American Section that this refusal was part of a general policy to conserve US currency.[60] Even when a British colony would authorise imports from Puerto Rico, the workings of imperial preference could make them expensive. According to Moscoso, preferential tariffs were adding around 10 per cent to goods from Puerto Rico so that when it came to the Barbados cement market, British cement retailed at 55 cents per barrel while Puerto Rican cement was $1.28 per barrel.[61] To make matters worse, new controls were introduced after the sterling crisis of 1947. Puerto Rico found that when it sought to import coconuts from British territories for a desiccating plant, a new Oils and Fats Agreement of 1947 allowed copra and copra products to be sent out of the British colonies only if demand within the British Caribbean had been satisfied.[62]

Aside from making little progress in persuading British territories to reduce discriminatory trade barriers, the US Section also found that its attempts to coordinate industrial development across the Caribbean region came to nothing.

Again it was Moscoso who brought Puerto Rico's grievances to the Commission when he complained about a glass factory that had been established in Trinidad by the Caribbean Development Corporation. The Caribbean Development Corporation was a private company operating in Trinidad that owned a brewery and was expanding into related industries such as bottle manufacturing. In a planned regional economy, the government of Trinidad would not have allowed a new business to be established in competition with an existing one and would have bought its glass from Puerto Rico where a glass factory had been operating since 1942.[63] Britain, however, seemed only interested in aiding the development of industry in British colonies.

The most concerted attempt to introduce machinery for the wider coordination of industrial development across the Caribbean came with the industrial survey of 1947–48. A panel of four experts representing the US, Britain, France and the Netherlands made industrial surveys in their territories and then a French economist, Luc Fauvel, prepared the final report. The two main proposals that emerged were the creation of a Caribbean Economic Policy Committee that would formulate a general economic policy for the region and an American-backed Caribbean Bank that would provide loans to colonies who adhered to that policy.[64] Fauvel reported a strong desire on the part of the expert from the US Section, the Puerto Rican official Rafael Picó, to see more action to stimulate industrial development.[65] Picó, of the Planning, Urbanizing and Zoning Board of Puerto Rico, was one of a number of Puerto Ricans who were appointed to the US Section of the Caribbean Commission; the other high-ranking Puerto Rican official that served in 1949 was the former Governor, Jesus T. Piñero. These Puerto Rican members of the US Section were more vocal than other officials in agitating for greater government intervention to foster rapid change in the conditions of the Caribbean colonies, both to alleviate the plight of poor Caribbean peoples but also as regional development was of benefit to the Puerto Rican economy.

The suggestions made by Fauvel and supported by Picó were rejected, almost in their entirety, by the British expert, Robert Galletti, who was a constant dissenting voice in the ensuing discussion. Galletti had been chosen by the Colonial Office to represent the British view. He had a degree in economics and had been an officer in the Indian Civil Service where he was Joint Secretary of the Board of Revenue of Madras. While little information is available on Galletti, his views on industrialisation show him to be an ardent believer in the free play of market forces. Galletti told the other experts from the Industry Survey that in his view any attempt to enforce Caribbean-wide policy would be construed by industrialists as an attempt to limit their freedom of action and would deter them from investing in the region:

> The highly individualist business man of the West Indies certainly does not desire to have his development 'co-ordinated' by any authority able to prevent him from using his own judgment, making his own mistakes and reaping his own rewards.[66]

Similarly, Galletti stated that the governments of the British Caribbean colonies were unlikely to agree to the establishment of any coordinating body that could force a policy on them. Galletti's political stance was revealed even further when he referred with disdain to an attempt to create a 'Gosplan': a Soviet-style central body that planned and directed industrialisation. For Galletti, economic development was not compatible with overt centralised planning and the role of government was to guarantee freedom of action for British investors and colonial governments. Galletti rejected the proposed industrial development bank as unnecessary on the basis that Barclays Bank and the Colonial Development Corporation already gave loans to private individuals, public bodies or colonial governments. (In reality, the activities of the CDC in the British Caribbean had been limited, and privately the Colonial Office noted that the CDC only dealt with large projects and was not acting as a source of a capital for entrepreneurs hoping to start up small factories.)[67] The proposals that Galletti made instead were that the Caribbean Commission undertake a market research survey to provide information for businessmen considering investing in the region. He also recommended the Commission encourage the funding of scientific research on the lines of the sugar research that was done at the Imperial College of Tropical Agriculture in Trinidad funded by the CPRC.[68]

Aside from rejecting the proposals that came out of the industrial survey of 1948 on the grounds of political economy, the Colonial Office also expressed its concern about the political implications of the proposals set out by Fauvel, describing them as 'controversial and far-reaching'.[69] The new instruments of the Caribbean Commission recommended by Fauvel and the US Section threatened to give the Commission the power to dictate policy for the British colonies, circumventing the authority of Britain and undermining its prestige in the region and beyond. The desire of the American Section to see an expansion of the powers of the Caribbean Commission was a recurring issue for Britain. From its inception, the Americans were considered to regard the Commission as 'bigger than we do', in the words of the Colonial Office, and had a tendency to attempt to endow the body with executive powers.[70] The Colonial Office was intent on protecting the sovereignty of the Colonial Empire by limiting the function of the Commission to that of a body that collated information on social and economic matters.[71]

Nothing came of the proposals made by Fauvel. J. E. Heesterman in his role as Industrial Advisor to the Commission claimed that by the time the final text of Fauvel's report had been submitted to the Commission, a period of two years had elapsed and it was no longer considered up to date, and it is possible that the stalling of the British Section played a role in retarding any final decisions.[72] In 1949 Lawrence Cramer, who had replaced Taussig as the Secretary General of the Caribbean Commission, made another attempt to instigate some action by suggesting that experts be appointed again to collate information on industrial and agricultural development across the region. The reaction from some officials at the Colonial Office this time was hostile. W. D. Sweaney of the West Indies Department wrote in a note for his colleagues that the recommendation of Cramer would mean a drain on money and personnel in colonial governments, not only in producing the information requested but also to read 'the voluminous publications which the Caribbean Commission will produce. My impression is that the Caribbean Commission is already producing a great deal of unwanted and unread material.'[73] Sweaney complained that the production of yet more data would do nothing to attract any entrepreneurs to the region. Colonial Office exasperation with the constant fact-gathering and report-writing activities of the Caribbean Commission was underpinned by the belief that the CDW Org provided the British colonies with all the advice they needed. British officials complained that the US appeared unaware of how much technical advice was available to Britain's colonies and was ignorant of the improvements that Britain had engendered through scientific and medical work in its colonies. W. A. C. Mathieson, a member of the UK delegation to the UN, proposed a vigorous campaign of propaganda in 1951 to make the point 'that we virtually invented technical assistance'. Mathieson singled out one area in particular in which Britain was an acknowledged world leader, the field of sugar research.[74]

With the creation of Truman's Point Four programme and the emergence of international bodies such as the World Health Organization, new sources of technical assistance were being pushed on the British Empire. The Colonial Office did not deny that British West Indian colonies might wish to apply for foreign or UN funds or assistance and agreed to this, as long as London and the CDW Org were informed of all requests.[75] In private, officials were explicit, however, that everything should be done to encourage the colonies to continue seeking 'men, money and advice' from Britain. As Britain's colonies gained greater autonomy, it was clear that they might seek help from other sources, but

> It is important for the future of the Commonwealth and the links between them and us that they should do so as little as possible. It is the job of the Colonial Office to see that the transition to greater political autonomy and independence is made without prejudice to the friendship which we hope to maintain indefinitely with those who are now 'the Colonial peoples'. The scientific help and guidance which this country gives to the solution to the many problems facing colonial peoples is and can remain a powerful force for the maintenance of such friendship.[76]

The provision of scientific advice was a key strategy by which Britain hoped to maintain a relationship with its former colonial territories, and from this perspective, American technical assistance was unwelcome.

Advisors from the Caribbean Commission were aware that they were involved in competition with experts deployed directly by the Colonial Office and CDW Org. During the 1940s, Britain had the advantage when it came to the take-up of advice by the British Caribbean colonies to the extent that by the early 1950s, some officials at the Caribbean Commission were complaining that they were not making any impression on Caribbean politicians and policy makers. C. J. Burgess, Executive Secretary of the Economics Section of the Central Secretariat, wrote to Eric Williams in 1951, 'There can be no doubt that the work of the Commission has suffered severely from the isolation of its staff from the territories and the failure to educate people in the area to its purposes and potentialities.'[77] Burgess stated that the British staff of the CDW Org had a far higher profile in the Caribbean territories and he credited this to the fact that British advisors were more active in getting out to the colonies and proffering advice in person to colonial governments while Caribbean Commission staff stayed behind their desks. Burgess was inspired to undertake a tour of the Caribbean to assemble data for his own Industrial Development Survey.[78] This document was presented at a major conference held by the Caribbean Commission in 1952. This meeting represented the next attempt by the Commission to take a leading role in determining the direction of policy for industrialisation in the region by providing expert reports and recommendations.

W. A. Lewis and Caribbean industrialisation

The various attempts made by the Caribbean Commission during the 1940s to formulate and implement coordinated Caribbean-wide policy for industrial development produced very little, due in no small part to resistance from Britain. Despite its poor record of success, the Caribbean Commission pushed for ways to coordinate policy for industrial development and force a greater pace of change.[79] The commissioners from Puerto Rico on the US Section of the Commission continued to promote their development programme as a model. The value of Operation Bootstrap as a template for industrialisation received an important endorsement when W. A. Lewis was asked to investigate the programme and then used it as the basis of a comprehensive vision of industrialisation specifically for the British West Indies. This was particularly problematic from the Colonial Office perspective as he did so on the request of Eric Williams, the Trinidadian academic who was Deputy Chairman of the Caribbean Research Committee and a member of the British Section of the Caribbean Commission.[80] Whether Williams invited Lewis to make a report with the clear intention of subverting the policy promoted by the Colonial Office cannot be said for certain, but Williams had a history of using his position on the Commission to criticise British rule, and his relationship with the Colonial Office could be fractious.[81] Williams eventually resigned from the British Section of the Commission in June 1955.

In 1951, Williams made the suggestion that Lewis (now Stanley Jevons Professor of Economics at the University of Manchester) be engaged as a consultant to the Commission to produce criteria that could be used to judge the suitability of potential Caribbean industries. Lewis accepted and undertook a tour of Trinidad, Puerto Rico, Jamaica and British Guiana before producing two articles in the

Caribbean Economic Review: 'Industrial Development in Puerto Rico' in 1949 and 'The Industrialisation of the British West Indies' in 1950.[82] He presented his studies at the Caribbean Commission's Industrial Development Conference in Puerto Rico, 11–20 February 1952. Officials at the Colonial Office found that Lewis's opinions on the best ways to encourage industrialisation, views that they had previously rejected when Lewis had been a member of CEAC, now had a new and high-profile outlet. Even worse, Lewis was using the platform that had been given to him by the Caribbean Commission to make direct criticism of Britain's colonial policy.

Puerto Rico was chosen as a venue for the conference so that delegates could see Operation Bootstrap for themselves.[83] The industrialisation programme begun in 1942 had expanded a great deal. After beginning with government-run factories of questionable value, the scheme had switched to a policy of attracting outside investment. In 1948 a law was passed giving tax exemption for ten years for investors creating factories on the island. In 1945 PRIDCO established an office in New York and promotional offices were subsequently created in Miami, Los Angeles and Chicago that worked to attract American business to Puerto Rico. The Public Relations department placed articles in the press, published pamphlets, staged exhibitions and produced films. Moscoso engaged the services of the advertising executive David Ogilvy, who devised a campaign that ran in American newspapers and magazines. The campaign aimed to change the image of Puerto Rico as a backward island and instead sell it as an economically attractive and politically stable place for American business to set up factories. Puerto Rico was promoted as the only part of the US where industry could operate with complete tax exemption.[84] In addition, the island offered the benefits of relatively low wages for labour, and since Puerto Rico was considered part of America, the territory used the US dollar as currency and there were no import duties for goods shipped to the mainland. Other incentives provided by PRIDCO, and then the Economic Development Administration formed in 1950, included the acquisition of land and construction of factory buildings, subsidised rents on industrial premises and loans to industry.[85] By creating conditions perceived favourable to private industrial capital, Puerto Rico saw twenty-four new industries established by 1948 and over 300 by 1955.[86] This 'industrialisation-by-invitation' approach was promoted as a model for other developing areas by both Puerto Rican members of the US Section and also officials from the US mainland, such as the Secretary General of the Commission, Lawrence Cramer, who reported on the programme for the Economic and Social Council of the UN.

In 'The Industrialization of the British West Indies', Lewis stated that the Puerto Rican experience demonstrated 'why the industrialisation of a new country cannot just be left to the ordinary forces of the market, but demands very positive and very intelligent action by governments'. This was a rebuke to the Colonial Office and the approach advocated by Caine. Lewis criticised the industrial policy of the Colonial Office for its adherence to laissez-faire economics:

> The basis of the laissez-faire philosophy is simply the belief that, if anything is worth doing, then someone will do it. If no one does it, then it cannot be worth doing, and the effort of a government to get it done must be contrary to the public interest. On this view, it is not necessary for a government specially to promote industrialisation, for, if industries are worth establishing, then private persons will establish them. The sphere of the government is confined to helping in the usual ways, such as paying for technical education, or maintaining communications.[87]

Instead, Lewis argued for various forms of government assistance. As in Puerto Rico, the other territories of the Caribbean needed to focus their efforts on attracting outside investment. A customs union created by the federation of British territories was advisable to create larger regional markets for prospective industries and to prevent competing industries emerging in the colonies of the Caribbean. In addition, the British West Indies needed to attract British and American manufacturers who exported to Latin American markets to set up their factories in the British West Indies. This was not an easy matter as industrialists would always prefer to go to a place that was already developed. Special measures were needed to attract the first pioneer industries, including tax holidays, monopoly rights, subsidies or tariff protection, and a period of 'wooing and fawning'.[88]

Lewis's recommendation was for the creation of an industrial development corporation to serve the whole British West Indies. Its job would be to decide on the types of industry that might be useful, to advise government on the assistance to be offered to new industry and undertake public relations work to interest manufacturers.[89] In a reiteration of the vision of development that he had put to the Colonial Office almost ten years earlier, Lewis said that the corporation should focus on creating trading estates with electric power and transport connections, so that factories were concentrated in one area to reduce their costs. It would also be useful if the industrial development corporation built factories to be leased, as the Labour Government had in fact done in Britain's depressed areas at home. Modelled on the Industrial Development Company in Puerto Rico, the industrial development corporation would have offices in London and New York in order to make the contacts necessary to attract business to the Caribbean. It would require a substantial budget and a large and expert staff based in the UK and US to advertise the benefits of the British West Indies.[90] Finally, when it came to sources of finance available to fund new industry, Lewis rejected the suggestion that the CDC was sufficient, stating that the CDC did not represent the interests of the West Indies but was 'a creature of the United Kingdom government'. Although attracting foreign capital was the priority, Lewis said it was probably necessary for the British Caribbean to have an industrial development bank to act as a lender of last resort.

Over the next decade, some of Lewis's recommendations exerted much influence on the policies for industrial development adopted by Britain's Caribbean territories. In the short term, however, the Colonial Office rejected the recommendation for the creation of an industrial development corporation on the basis of cost.[91] The Colonial Office also doubted, incorrectly as it turned out, that any colonial government would be impressed by what officials described as the low living standards of Puerto Rico and rush to embrace Lewis's suggestions.[92] British officials continued to adhere to a vision of industrial development for the Caribbean that conceded there was a need for government action to encourage industrialisation but wished to restrict this role to the provision of services and information with some short-term forms of tax relief.

Conclusion

The 1940s were a time of unprecedented interest in the encouragement of industry in places traditionally reliant on an agricultural sector, many of which had suffered badly during the Great Depression. Discussions at the metropolitan, imperial and regional level during the 1940s and 1950s led to the emergence of a number of proposals for the industrial development of Caribbean region. Whilst Colonial Office policy marked a break with the past in embracing the need for some degree of industrial development, it was cautious and limited in contrast to the ideas of the architects of Puerto Rico's industrialisation programme or the proposals of Lewis. When left-wing economists and proponents of the New Deal considered the future of the Caribbean, they prioritised government-led change and focused on the need for transformation of colonial economies to improve living standards, help invigorate a regional and world economy and demonstrate the altruism and efficacy of American-backed ventures operating after 1945.

While the Colonial Office spoke of the need for greater metropolitan action to encourage development after 1940, it avoided making large financial commitments. The desire to avoid tying colonial and metropolitan governments into long-term financial responsibilities can be seen to have stemmed in part from awareness that the support that came from the British government for development had limits. Britain found itself in straitened financial circumstances in the post-war world, in contrast, of course, to the US. The economic crisis of 1947 caused a reordering of priorities at the Colonial Office. The self-interest of Britain was asserted over the needs of the colonies and there was focus on increasing production of primary products in the colonies that might earn dollars or alleviate shortages at home. Apart from this, however, the principles upheld by the Economics Department of the Colonial Office reflected long-held beliefs amongst officials on what constituted sound political economy. Lewis described Caine's department as 'the last refuge in this country of what is popularly called 19th century laissez-faire'.[93]

A lack of enthusiasm for institutions such as the government-run industrial development corporations and publicly funded development banks may well have contributed to the relative invisibility of the British approach, both at the time, and to historians subsequently. One area in which the Colonial Office did encourage state intervention to facilitate industrialisation was through the funding of scientific research, a priority unique to the Colonial Office. Under the CPRC, two new scientific laboratories were created in Trinidad to carry out work that was anticipated to be the basis of industrial diversification. These laboratories were intended to be highly visible symbols of the modernising intentions of Britain. They also represented a resolution of the question of how government could intervene to stimulate economic change without compromising the tenets of liberal political economy.

Notes

1 W. A. Lewis, "Economic development with unlimited supplies of labour", *The Manchester School* 22 (1954), 139–191.
2 Lee and Petter, *The Colonial Office, War and Development Policy*, pp. 171 and 172.
3 TNA, CO 852/577/1.
4 Butler, *Industrialisation*, p. 38.
5 TNA, CAB 24/249.
6 Butler, *Industrialisation*, p. 38.
7 TNA, CO 852/577/1.
8 B. Ingham, "Shaping opinion on development policy: economists at the Colonial Office during World War II", *History of Political Economy* 24 (1992), 689–710.
9 Butler, *Industrialisation*, p. 91; M. Petter, "Sir Sydney Caine and the Colonial Office in the Second World War: a career in the making", *Canadian Journal of History* 16 (1981), 68–85.
10 TNA, CO 852/588/2.

11 *Ibid.*
12 *Ibid.*
13 C. Whitham, *Bitter Rehearsal: British and American Planning for a Post-War West Indies* (Westport: Praeger, 2002), p. 38; C. Fraser, *Ambivalent Anti-Colonialism: The United States and the Genesis of West Indian Independence, 1940–1964* (Westport: Greenwood Press, 1994), pp. 59 and 64.
14 Parker, *Brother's Keeper.*
15 Whitham, *Bitter Rehearsal*, p. 113; National Records and Archives Administration (NARA), College Park, Maryland, USA, RG 43 UD 07D 82, Box 3 A1–5c.
16 J. Parker, "'Capital of the Caribbean': the African American–West Indian 'Harlem Nexus' and the transnational drive for black freedom, 1940–1948", *The Journal of African American History* 89 (2004), 98–117.
17 Parker, *Brother's Keeper*; Whitham, *Bitter Rehearsal*, pp. 38–45; Fraser, *Ambivalent Anti-Colonialism*, pp. 48–49.
18 C. Taussig, "A program in the Caribbean", *Foreign Affairs* 24 (1946), p. 703.
19 *Ibid.*
20 TNA, CO 1042/14.
21 A. W. Maldonado, *Teodoro Moscoso and Puerto Rico's Operation Bootstrap* (Gainesville: University Press of Florida, 1997), p. 13.
22 TNA, CO 1042/14.
23 TNA, CO 990/17.
24 Maldonado, *Teodoro Moscoso*, ch. 4.
25 Maldonado, *Teodoro Moscoso*, p. 40.
26 Maldonado, *Teodoro Moscoso*, p. 42.
27 Maldonado, *Teodoro Moscoso*, p. 44.
28 NARA, RG 43 UD 07D 82, Box 2.
29 Butler, *Industrialisation*, pp. 174–181.
30 Butler, *Industrialisation*, pp. 188–189.
31 TNA, CO 1042/14.
32 TNA, CO 314/452/2.
33 TNA, CO 852/577/1.
34 TNA, CO 852/577/1.
35 TNA, CO 1042/14.
36 TNA, CO 1042/14
37 TNA, CO 852/577/1/.
38 Butler, *Industrialisation*, pp. 102–108.
39 TNA, CO 990/1.
40 TNA, CO 990/17.
41 TNA, CO 990/17.
42 TNA, CO 852/588/2.
43 TNA, CO 852/588/2.
44 TNA, CO 990/17.
45 Lewis Papers, Princeton University, Box 26, Folder 11.
46 Lewis Papers Princeton University, Box 26, Folder 11.
47 TNA, CO 1042/151.
48 TNA, CO 1042/151.
49 TNA, CO 1042/14.
50 TNA, CO 852/578/3.
51 TNA, CO 852/578/3.
52 TNA, CO 852/578/3.
53 TNA, CO 852/578/3.
54 TNA, CO 318/501/5.
55 NARA, RG 469, Box 18.
56 NARA, RG 43 UD 07D 82, Box 1.
57 NARA, RG 43 UD 07D 82, Box 23.
58 NARA, RG 43 UD 07D 82, Box 2 A1–4.
59 NARA, RG 43 UD 07D 82, Box 29.

60 NARA, RG 43 UD 07D 82, Box 29.
61 NARA, RG 43 UD 07D 82, Box 29.
62 NARA, RG 43 UD 07D 82, Box 39, 2–5(a).
63 NARA, RG 43 UD 07D 82, Box 6.
64 TNA, CO 852/1052/2.
65 TNA, CO 852/1052/2.
66 TNA, CO 852/1052/2.
67 TNA, CO 1042/152.
68 TNA, CO 852/1052/2.
69 TNA, CO 852/1052/2.
70 TNA, CO 314/452/22.
71 TNA, CO 318/480/4.
72 TNA, CO 318/501/6.
73 TNA, CO 318/480/3.
74 TNA, CO 318/480/3.
75 TNA, CO 852/1321/4.
76 TNA, CO 900/13, memorandum December 1954.
77 Eric Williams Memorial Collection, University of the West Indies, Trinidad (EWMC), Box 044.
78 EWMC, University of the West Indies, Trinidad, Box 044.
79 TNA, CO 318/501/6.
80 EWMC, Box 063.
81 TNA, CO 1042/68.
82 L. W. Cramer, Foreword, *Industrial Development in the Caribbean* (Port-of-Spain, Trinidad: Caribbean Commission Central Secretariat, Kent House, 1951).
83 TNA, CO 318/501/6.
84 Maldonado, *Teodoro Moscoso*, pp. 81–84.
85 E. Pantojas-Garcia, *Development Strategies as Ideology: Puerto Rico's Export-Led Industrialisation Experience* (Boulder: Lynne Rienner, 1990), pp. 78–81.
86 Pantojas-Garcia, *Development Strategies*, pp. 78–81.
87 W. A. Lewis, "The industrialization of the British West Indies", in *Industrial Development in the Caribbean* (Port-of-Spain, Trinidad: Caribbean Commission Central Secretariat, Kent House, 1951), p. 54.
88 Lewis, "The industrialization of the British West Indies", p. 56.
89 Lewis, "The industrialization of the British West Indies", p. 58.
90 Lewis, "The industrialization of the British West Indies", p. 55.
91 TNA, CO 318/501/6.
92 TNA, CO 318/501/6.
93 Arthur Lewis, "Colonial Development", Lecture to Manchester Statistical Society, 12 January 1949.

CHAPTER FOUR

Laboratory science, laissez-faire economics and modernity

The work of the CPRC to identify new uses for sugar was incorporated into Colonial Office plans to encourage industrial development in Britain's Caribbean colonies. Expanding on its role as a sponsor of research at British universities, the CPRC created a new laboratory for sugar research in Trinidad in 1951 with the goal of inspiring West Indian sugar producers to diversify their interests and establish chemical factories in the Caribbean. A second laboratory was created in Trinidad to carry out research into microbiological problems related to manufacturing, medicine and agriculture. The Colonial Microbiological Research Institute (CMRI) was the only institute for tropical microbiology in the British Colonial Empire. The two laboratories in Trinidad were intended to be a tangible, visible intervention to cultivate industrial development in the British Caribbean at a point when the Colonial Office mostly offered advice. The debates of the 1940s on the best way to encourage economic diversification revealed a tendency amongst British officials to discourage the adoption of measures that were too protectionist in nature or that might incur a financial burden for colonial consumers and British taxpayers. The state funding of scientific research to identify industrial uses for sugar, however, represented a resolution of the issue of how to take some action to encourage industry whilst still adhering to laissez-faire principles. The two laboratories created in Trinidad were described as places of fundamental research, and this designation worked to carefully distinguish actions undertaken by the state in the name of development from activities that were considered the business of the firm. State-funded fundamental research provided basic information about the chemistry of sugar. It was left up to business to exploit this information in order to develop products for the market. Thus, the Trinidad laboratories were a contribution to a mode of industrial development that was essentially liberal in tenor.

The use of such careful distinctions between the role of the state and the remit of the firm shows that an emphasis on planning and information gathering, and the rise of experts, do not necessarily result in highly authoritarian, centralised and far-reaching systems of state-controlled development. The path to economic development envisaged by the CPRC and the Colonial Office had at its heart the preservation of freedom for both scientists and entrepreneurs to make their own decisions. This development vision for the British West Indies sits in contrast to development schemes where state experts oversaw an attempt at the wholesale remaking of African societies and their economic activities into something susceptible to control and supervision.[1] This extreme reordering of nature and society by the state in the name of development was not a feature of British economic diversification plans for the British Caribbean. In this case, the links between scientific research and economic activity were loosely coordinated. It was a form of development where the initiative in establishing new industry ultimately lay with businessmen who would make judgements about compounds offered up by science.

While representing a more liberal version of development than the large-scale agricultural projects in Britain's African colonies, the interventions that occurred under the aegis of the CPRC were still intended to be highly visible representations of imperial power; concrete manifestations of Britain's commitment to modernising its colonial possessions. A vision of modernity for Trinidad was expressed through the construction of chemical and microbiological laboratories furnished with the most up-to-date equipment and, importantly, which functioned as world centres. The CMRI in particular was expected to be a symbol of nascent Trinidadian modernity as it endowed the island with the ability to participate in the international project of scientific advance, something that a study of strictly local problems would not have done. The general nature of fundamental research in Trinidad – the fact that it investigated basic underlying laws and processes that operated in all locations – was crucial in bestowing this greater prestige. The Colonial Office celebrated the ability of Trinidad's new laboratories to allow the island to participate in the global circulation of knowledge as evidence of Britain's sincerity of purpose in developing and modernising its colonies.

Industrial development and research into colonial products

At the very first meeting of the CPRC in January 1943 the committee were told that the Secretary of State sought the establishment of new processing industries that were located in the colonies themselves as part of the drive towards economic diversification. This goal of encouraging colonial industry became part of the general policy of the CPRC, as stated in their first report:

> The low standard of living for Colonial peoples is, at any rate in part, due to the fact that they are almost without exception primary producers and therefore do not enjoy the higher standards which can, generally speaking, be attained only by industrial activity. It must therefore be constantly alive to the possibility of developing techniques whereby Colonial peoples may not only produce primary products but also convert them into secondary products of greater value, both for internal consumption and for export.[2]

The Colonial Office made clear to the CPRC that government was not expecting to operate new factories itself.[3] The vision of industrial development that was to be employed was one where scientists funded by the council might discover new products or processes and then business would take them up. In the case of the British West Indies, factories were sought that would produce secondary products for export, in contrast to the import-substitution industries that were anticipated for Britain's African colonies.[4] The question for the CPRC was how to persuade manufacturers to take up new compounds discovered through scientific research and establish factories in the British Caribbean.

The relocation of industry producing industrial alcohol from Britain to the British West Indies seemed an obvious way to establish new colonial factories. Before the outbreak of the Second World War, molasses produced in the Caribbean was shipped to Britain by the United Molasses Company (UM) in its fleet of tankers, the Athel Line, before distillation into alcohol by the Distillers Company Ltd (DCL).[5] The relationship between UM and DCL was a close one and they exchanged shares and directors in 1930. DCL controlled over 80 per cent of the supply of alcohol to industry in Britain and its customers converted industrial alcohol into an enormous range of chemical products, including synthetic plastics and fabrics. The large size of the market for industrial alcohol inspired the Director of Research of the CPRC, John Simonsen, to see if British firms such as DCL and ICI might move key processes in which sugar and molasses were converted into alcohol from Britain to the British West Indies.

Simonsen decided to try to persuade officials and economic advisors at the Colonial Office that it was in Britain's best interests to explore the establishment of industry that converted plant products into important chemical intermediates in the colonies. He argued that Britain, unlike the US, did not have access to a large supply of oil that could be used as a raw material for synthetic manufacturing. The prospect of Britain trailing behind the US in terms of synthetic chemistry had been invoked by the CPRC on more than one occasion, and the idea that the CPRC had a vital function in enabling Britain to catch up provided an important rationale for the work of the committee. Simonsen informed the Colonial Office that in the absence of cheap oil, the British chemical industry would have to rely on plant products such as sugar and cellulose. He claimed that butylene glycol and rubber could be made more cheaply from sugar than from petroleum, and made the bold claim that 'Most chemists thought that the future of these industries depended on using plant products.'[6] Government was needed to act 'as a catalyst', encouraging companies such as ICI, DCL, and Kemball and Bishop to collaborate in establishing overseas factories based on tropical products.[7] Trinidad was the most likely location for new manufacturing as it had the only oil industry in the empire and therefore it had chemists, engineering facilities, tank storage for molasses, and 'a certain industrial atmosphere'.[8]

Consultation between the Colonial Office and representatives of Kemball and Bishop and ICI revealed, however, that these two companies had no interest in setting up factories in Trinidad. ICI suggested instead that the Colonial Office might find colonial sugar manufacturers interested in producing industrial alcohol on a large scale in the British West Indies, a suggestion received with enthusiasm by officials. ICI alerted the Colonial Office to one factor that would be important in terms of cost – the duty currently paid on imports of alcohol.[9] Domestically produced alcohol was subject to an inconvenience allowance that made alcohol manufactured from molasses in Britain cheaper than coal or oil as a raw material for firms such as ICI. In January 1945, however, Customs and Excise informed the Colonial Office in a confidential letter that the inconvenience allowance was going to be withdrawn later that year. Since the relative low cost of domestic alcohol had previously acted as a disincentive to colonial production, this was news that appeared to improve the prospects of an industry based in the British West Indies that produced alcohol for import to the UK.

The optimism of officials was suddenly punctured, however, with the revelation in December 1945 that ICI had developed a cheap process for producing synthetic alcohol from ethylene. Ethylene gases were a product of oil, and this new process signalled a move by ICI towards the use of oil as a starting compound for manufacturing its synthetic products. This news threatened to make the colonial production of alcohol by fermentation of molasses or sugar entirely redundant.[10] Ethylene from oil was well established as a source of alcohol in the

US, produced by companies such as the Carbide & Carbon Chemicals Corporation, a subsidiary of Union Carbide. Ethylene had not been adopted by British chemical companies before the Second World War because the inconvenience allowance made alcohol from molasses a cheaper starting material.[11]

A meeting with Michael Kielberg, Chairman of UM, in January 1946 did not give much further encouragement to the idea that a Trinidad-based alcohol industry was feasible. Kielberg expressed the view that a colonial alcohol industry would struggle to be economic. The distilleries in the region would be small, and bringing molasses from other islands to Trinidad would be costly, even more so when losses due to evaporation were included. Kielberg told the Colonial Office that the cost of transporting molasses from other Caribbean colonies to Trinidad during the war had been as much as the freight from Trinidad to Britain. In addition, UM had found that alcohol shipped to Britain needed to be re-distilled on arrival, however clean the tanker. Sir William Rook, the Director of Sugar at the Ministry of Food, described Kielberg to the Colonial Office as trustworthy, discreet and reliable, and as far as many officials were concerned, his opinion on the unlikely future for colonial alcohol for export brought the matter to an end.[12]

Simonsen failed to persuade British chemical firms to create a fermentation industry based in the West Indies, and the fact that ICI was now seriously exploring oil as a starting material raised questions about the future prospects of existing companies that produced industrial alcohol from molasses. Simonsen was reluctant, however, to abandon the idea of new alcohol plant in the colonies altogether, criticising what he described as the 'defeatist tone' of the Colonial Office. He argued that establishment of an alcohol industry would create a nucleus for other chemical manufacturing, and that rather than assessing its value to UK chemical firms, industrial alcohol production needed to be valued in terms of its contribution to the establishment of a range of new colonial industries.[13] Whilst conversations had been under way in London with British chemical companies, the CPRC had been exploring an alternative route to industrial diversification in the British West Indies that focused on new lines of production by Caribbean sugar manufacturers. In this vision of industrial development, the British West Indies sugar producers would collectively fund research into finding industrial uses for sugar cane and its by-products so they could diversify their products. The next step was to persuade sugar manufacturers that operated in the British West Indies to participate in such a scheme.

SCIENCE, LAISSEZ-FAIRE ECONOMICS AND MODERNITY

The British West Indies Sugar Research Association

Simonsen and Robert Robinson went in person to a meeting of sugar manufacturers in Barbados in August 1944 in order to try to persuade them to invest in sugar research as the first step towards diversifying their activities. The meeting was part of a longer tour that included visits to Canada and the US as well as the Caribbean. Robinson wished to meet with Trinidad Leaseholds, the producers of oil in Trinidad, to discuss his proposed research into chemical products derived from petroleum. This was a field that Robinson claimed it was necessary for the CPRC to fund so Britain could keep up with work into petroleum products that was under way in the United States.[14] Robinson also had reason to visit the US as he was engaged in the development of penicillin and the trip afforded him the opportunity to meet with American workers based at the Northern Regional Laboratory of the Bureau of Agriculture at Peoria, Illinois, who were working on the large-scale production of the drug. On their return, Simonsen and Robinson reported to the CPRC their admiration of the work at Peoria on large-scale fermentation processes, an area where yet again it was said Britain lagged behind the US.[15]

The group also visited the Imperial College of Tropical Agriculture (ICTA), where a Department of Sugar Technology had operated since 1924. Scientists based at the ICTA in Trinidad were unable to return home during the war and the Colonial Office believed they would appreciate a visit from academic colleagues from Britain. The ICTA in central Trinidad was one of only two major research centres in agriculture in existence in the Colonial Empire before the Second World War (the other was Amani in Tanganyika). Apart from carrying out research into soils, pests and the improvement of banana and cocoa, from 1927 the college had the important function of training all cadets entering the Colonial Agricultural Service. The ICTA had suffered from a lack of funding during the Great Depression, and staffing levels and the quality of equipment had further declined during the war. Simonsen and Robinson were not impressed by conditions at the college when they arrived in 1944. They reported back to London that researchers were overburdened by teaching and the laboratories were so inadequate that pairs of students worked with less room 'than would be given to a single boy in a British Secondary School'.[16] The visitors noted in particular that the equipment of the model sugar factory in the Department of Sugar Technology was very old and needed renewal.[17] This department trained most of the sugar chemists who worked in the firms that operated in the British West Indies.

On leaving Trinidad, Simonsen, Frank Stockdale and Robinson went to Barbados to meet with the British West Indies Sugar Association (BWISA) in order to try to persuade sugar manufacturers to make a financial contribution to scientific research into sugar. The BWISA was formed in 1942 by the directors of the associations of sugar producers in Antigua, Barbados, British Guiana, Jamaica, St Kitts, St Lucia and Trinidad. Its main purpose was to lobby the British government on issues such as price agreements and quotas.[18] In his speech to the sugar manufacturers at the meeting in August 1944, Simonsen described how the CPRC had been formed under the 1940 CDW Act and how it currently supported research into tropical products in university laboratories in Britain. He told his audience that the CPRC intended 'to raise the standard of living within the various colonies and to increase their prosperity' through research into the use of sugar as a raw material for the chemical industry. Sugar, he said, was 'the purest organic chemical known'.[19] Robinson then suggested that BWISA could create a sugar research association on the lines of the research associations that had been created under the DSIR, with the British government providing 50 per cent of the funding for research from the Research Fund of the 1940 CDW Act.[20]

The DSIR had started a research association scheme (referred to as the Million Scheme) in 1917 with the aim of encouraging firms in Britain to invest in scientific research. The expectation was that participation in the research association scheme would demonstrate to industry the value of scientific research, and firms would then independently support this work, maybe through the formation of their own laboratory. From the original Million Scheme twenty cooperative research associations were in operation by 1932, including the British Photographic Research Association, the British Portland Cement Research Association and the British Research Association for Cocoa, Chocolate, Sugar, Confectionery and Jam Trades, and the scheme was subsequently expanded so that forty associations were in existence by 1944. Firms in a research association pooled funds to support research in a university or institute and government contributed a pound for every pound from industry. The results of research were freely available to all the participating firms.[21] Simonsen hoped that a British West Indies Sugar Research Association comprised of all the major sugar companies in the British Caribbean would be able to fund, on a 50:50 basis with the Colonial Office, a laboratory specially created to research industrial uses of sugar and located at the ICTA. It was put to the BWISA that scientific research into industrial uses of sugar and its by-products, 'consumption research', was a relatively neglected area in comparison to sugar cane breeding or the improvement of the milling process.[22]

The BWISA rejected the suggestion that they collaborate for the funding of research into new uses for sugar on the basis that if they were seen to be allocating money for this purpose, this would prompt sugar workers on the estates to ask for higher wages.[23] The unwillingness of the industry to make a new financial commitment was also due to uncertainty about the future. The British government had pledged to purchase the entire exportable surplus of sugar from the West Indies until 1946, but it was not clear what would happen after this. Quotas and prices had not been agreed and the International Sugar Agreement that had previously brought some stability to the price of sugar had been suspended with the outbreak of the war. The BWISA took the opportunity to argue that the price paid by Britain for colonial sugar was too low and made it impossible for the manufacturers to contemplate additional expenditure.[24]

The Colonial Office persisted with their proposals and sent a detailed report to the BWISA in October 1945 that attempted to explain the commercial potential of the new substances that had been discovered in Haworth's Birmingham laboratory. These included levulinic acid, which could be converted into an anti-freeze, and varnishes, plastics and sulphonamide drugs.[25] The Colonial Office warned that as the researchers at Birmingham discovered new products, there was always the danger that the development of these would be taken up by foreign firms. The creation of a sugar research association would ensure that British industry would be the main beneficiaries from the discoveries of British science. In addition, if sugar manufacturers diversified the uses to which sugar could be put this would afford some protection from any future uncertainties in the world market.[26] Eventually, the BWISA agreed to a cess of 4d per ton for research (approximately £10,000 a year), with another £10,000 coming from the Research Fund of the 1940 CDW Act.[27] The money was used to establish a sugar research scheme under Leslie Wiggins, inaugurated on 1 September 1947. Research work began in July 1948 at Birmingham University before the scheme was transferred to Trinidad to work in new facilities that had been built at the ICTA.

The Sugar Technology Laboratory (STL) was formally opened at the Silver Jubilee celebrations of the ICTA on Saturday 17 March 1951. The Francis Watts Laboratory, named after the first principal of ICTA, comprised a single-storey building in the grounds of ICTA with several laboratories and housing that could accommodate around twenty-six staff. Attached to the laboratories was a miniature factory for pilot trials of new products: the Experimental Sugar Factory. The capital for the building of these facilities came from CDW funds. Over 500 guests were invited to attend the jubilee celebrations of the ICTA, including

Figure 4 Aerial view of the Imperial College of Tropical Agriculture, c.1951. The experimental sugar factory is in the foreground.

previous principals of the college, the Governor of Trinidad, Sir Hubert Rance, Alan Burns of the Trusteeship Council of the UN, and the Bishop of Port-of-Spain. The *Trinidad Guardian* reported extensively on the celebrations, including coverage of the new sugar scheme.[28]

Simonsen read a message from Maurice Hankey, chair of the CPRC, at the opening ceremony that placed the new laboratory firmly in the context of wider aspirations for industrial development in the Caribbean. Hankey's text told the audience that the outcome of the work at the laboratory would be new industries in the West Indies that used sugar as their raw material so that 'based upon fundamental research and upon its application, and given reasonably stable political and economic conditions, we can anticipate an increase to prosperity in the West Indies.'[29] In his own speech, Simonsen explained that the new laboratory was modelled on those administered through the Research Association Scheme of the DSIR in Britain. Like these laboratories, the Francis Watts Laboratory was not intended to displace the day-to-day testing and ad hoc investigations that sugar firms did for themselves in their works laboratories. Instead, the Association's laboratory will only investigate major problems of interest to the whole industry. I will venture to suggest that this new laboratory will have two main functions (a) to carry out fundamental research on the utilisation of sugar and the by-products of the sugar industry, and (b) when desirable to carry out such investigations through the pilot plant stage. If these investigations show that the manufacture is likely to be economic and that a market for the product can be found then it will be for the Sugar Industry itself to undertake large scale production.[30]

By describing the work of the new laboratory as restricted to the investigation of only 'major problems of interest to the whole industry', Simonsen carefully demarcated the nature of its research. It would examine general issues, common to all firms, and not the problems specific to any individual business. This general research into the chemical utilisation of sugar and its by-products was labelled 'fundamental research'. By stating that the business of the research association was to sponsor fundamental research into major issues only, it was possible to avoid the suggestion that state funds for research would end up lining the pockets of a single firm. As fundamental research into sugar chemistry proceeded at ICTA, there would still be plenty of scope for investigations that dealt with issues specific to an individual company.

Simonsen's description of the function of the Francis Watts Laboratory closely resembled the words used by the DSIR when describing its own research associations that operated in Britain. In its early annual reports the DSIR had explained that there was more than one type of scientific activity that a firm could be involved in. While a business might need to undertake investigations of certain everyday problems in its works laboratory, there was also a need for study of the more general underlying issues that were common to all the businesses in the sector. The DSIR stated that this latter species of problem could be addressed by 'fundamental research'.[31] It was appropriate for government to make a contribution to general investigations or fundamental research as this would potentially benefit an entire sector of industry. The investigation of issues that were not broad or basic enough to be termed 'fundamental research' but were specific to the processes or output of one firm should not benefit from public funds. Government needed to avoid the implication that it favoured any individual company.

In the first half of the 1940s, officials in the Economics Department and members of CEAC such as W. Arthur Lewis had been engaged in fierce debate about the extent and the manner in which government should intervene to stimulate industrial development. For officials at the Colonial Office, using government funds for fundamental research into the chemistry of sugar provided an acceptable route for state intervention. In the public declarations that described the work of the sugar laboratory, the use of the term 'fundamental research' provided a rationale for state contributions to industrial research as a way of encouraging industrial progress. This rationale carefully delineated government funding of industrial research so that the involvement of the state in manufacturing did not compromise the tenets of liberal political economy.

Wiggins and his team pursued a number of lines of enquiry at the new laboratory in the grounds of the ICTA. These included the first in-depth study to identify the chemical constituents of cane juice and the production of itaconic acid by cultures of a fungus, *aspergillus terreus*, grown on a sugar medium. Itaconic acid could be used to make a Perspex-like plastic material, and investigations were done to find an economically viable method of large-scale production using the submerged culture methods that had yielded penicillin at Peoria in the USA.[32] In addition, the team of scientists in Trinidad explored production methods for compounds first developed at Birmingham University, such as furfural and levulinic acid. These compounds were versatile chemical intermediates that could be turned into a very large number of different products and therefore had the potential to be highly profitable. Furfural was used in the manufacture of nylon. Levulinic acid could be used to manufacture an effective anti-freeze and made a good carrier of calcium for use in medicine. In 1955 Wiggins reported that the laboratory was building a pilot plant with funds from an American chemical company involved in the plastics industry. The Argus Chemical Corporation was looking to produce levulinic acid on a commercial scale using a process developed by Wiggins and his team at the laboratory.[33] The firm came to an agreement with the sugar laboratory that on payment it would have the right to work the patent taken out by the BWISA in the USA while the BWISA retained the rights for the sterling area.

Much of the work done by the STL had a clear commercial orientation and Wiggins' team took out a number of patents, receiving advice on these from the National Research and Development Corporation during a visit to Britain in 1951. The STL investigated by-products of sugar processing such as sugar cane wax, molasses-based cattle feed, and the manufacturing of paper from bagasse, the fibrous part of the cane that was left after crushing had squeezed out the sugar-containing cane juice. Traditionally this waste product was used as a fuel in the estate factories for boiling sugar cane juice. The product that seemed to have the greatest commercial potential was sugar cane wax. Sugar cane wax could be used as a polish and was able to compete with carnauba wax in terms of price. The first commercial factory producing wax from cane had been established in Durban, South Africa, in 1916, and a factory making wax also operated in Australia. Interest in the development of sugar cane wax had been high in the US during the war because of shortages of other waxes, and research had been done in the Southern Regional Research Laboratory using Louisiana canes. The laboratory in Trinidad surveyed the yield of wax from sugar cane from across the British West Indies and found that Barbados cane had the highest wax content. By 1953, a factory had been established in Barbados producing sugar cane wax that received government support under the Pioneer Industries Legislation of Barbados, which gave tax relief to new business. The reports of the STL told how the wax was being sold locally in Trinidad and was used on floors and furniture at the San Fernando Hospital.

The focus on the development of compounds ready for the market shows clearly that the emphasis on 'fundamental research' in the original discussions of the CPRC and at the opening of the Trinidad laboratory did not mean that applied science and practical issues were to be neglected. Describing the work of the laboratory as 'fundamental research' allowed Wiggins the autonomy to pursue long-term and in-depth studies in organic chemistry if he wished. It did not restrict him to this work; it gave him the freedom to choose the sort of research his team would undertake with scope for more exploratory study if necessary, and in practice, the work of the Francis Watts Laboratory ranged from investigation of general phenomena to the resolution of manufacturing issues using pilot plant trials.

Apart from research, Wiggins and his team ran a three-year Diploma (DICTA) and a two-year postgraduate Associateship in Sugar Technology (AICTA). The Diploma recruited four or five students each year from sugar companies such as Booker Brothers Ltd. The AICTA attracted students from further afield, including two from Burma in 1956. It is not at all clear that these courses at the ICTA attracted any students born in the Caribbean. W. Arthur Lewis showed a great deal of disdain for the college, saying that ICTA had gone out of its way to avoid appointing British Caribbean scientists to senior posts and that students from the British West Indian colonies avoided going there as it had a reputation for racial prejudice. In his words, it was 'a little piece of England in Trinidad' whose primary function was to train colonial officers.[34] In the case of sugar research, Wiggins' department was mainly involved in training technical staff for the European-owned sugar companies.

The annual reports of the STL celebrated the incorporation of the work of the institute into an international network of researchers and

Figure 5 Staff of the ICTA, c.1951. Wiggins is seated in the front row, fourth from the left.

organisations concerned with the future utilisation of sugar. Wiggins had the task of advertising and promoting the work done by his team. In addition to his research work, he wrote articles on sugar cane as an alternative to coal and oil in journals such as *Chemistry and Industry*, and delivered lectures in Britain, including a talk to the Chemical Society at the University of Birmingham on 'Recent Advances in Carbohydrate Chemistry' in 1949.[35] The Sugar Foundation of New York, a trade association for American sugar manufacturers, awarded a grant of $15,000 to Wiggins and his team after a visit to the ICTA by 150 members of the International Society of Sugar Cane Technologists.[36] The Foundation went on to provide funds for a fellowship to study the reaction of molasses with ammonia with the aim of producing animal feed. A second grant to appoint a Research Fellow came from the Hawaiian Sugar Planters Association. This evidence of the international status of the laboratory was frequently emphasised in its annual reports. Wiggins wrote in 1953, 'It is gratifying to note that the sugar industry in other parts of the world has looked upon our efforts with favour and there is little doubt that the BWI Sugar Research Scheme has won a place in the forefront of the Sugar Research Institutes of the World.'[37] What was missing was any mention of the place of the laboratory with regard to the future of Trinidad. As with many of the forty other colonial research institutions that were created using funds from the 1940 and 1945 CDW Acts, the STL did not report to the colonial government in Trinidad but liaised directly with the Colonial Office, who oversaw appointments, grants and the research programme of the institute. Accordingly, the reports of the laboratory do not note any meetings between Wiggins and officers based in Trinidad who worked as part of the colonial administration, either in the central secretariat or officers from the agricultural or medical departments. The sugar laboratory was not expected to deal with issues that arose merely at the level of the individual colony, but instead its reach was regional and international. In the early 1940s the creation of new research bodies that had such a wide remit was considered essential to raise the quality of scientific research that occurred across the Colonial Empire. The STL afforded its scientists the chance to contribute to scientific understandings of the chemistry of sugar at the international level. This raised the prestige of the laboratory and was said to be important for recruiting the highest calibre scientists. In addition, the fact that the Colonial Office could describe Britain as a recognised world leader in the field of carbohydrate research was considered evidence of the extent of Britain's commitment to improving colonial conditions through technical interventions after 1940.

The first beneficiaries of the work done by Wiggins and his team were intended to be British or American firms. It fell to Wiggins to publicise the results of the research done at the laboratory to industry. This was a major task for a scientist who oversaw a research team of around twenty-six scientists and a programme of teaching. The annual reports published by the STL show that apart from giving lectures in Britain and writing articles, Wiggins met with representatives of British Caribbean sugar firms every year. He visited sugar estates, touring every factory in Trinidad during the 1953 crop, for example, and senior staff from companies such as Tate & Lyle visited the laboratory in turn.[38] Wiggins also spoke at meetings of sugar technologists beyond the British territories, going to Puerto Rico in November 1951 for a meeting with engineers, chemists and factory managers employed by Caribbean sugar manufacturers. In addition, the sugar companies Henckell du Buisson & Co Ltd and Booker Brothers seconded chemical engineers to work at the laboratory in 1954 and 1955. The staff of the Francis Watts Laboratory also had contacts with British chemical firms by virtue of the fact that some of its researchers had come to work there from industry – a physical chemist, W. S. Wise, joined from DCL in 1952. Beyond this, the Caribbean Commission helped to publicise the industrial applications of the work at ICTA, producing a lengthy document on the potential to create new Caribbean industry based on sugar products, *The Industrial Utilisation of Sugar-Cane By-Products*. This was an extensive and detailed guide to all the possible products that could be manufactured from sugar cane, prepared by a British consultant sugar technologist, Walter Scott. Scott celebrated the encouraging results coming out of Wiggins' laboratory in his introduction and recommended that any industrialist seeking to set up a business in the Caribbean would benefit from consulting the laboratory at ICTA. Scott's conclusion was there was a great deal of potential for establishing new factories in the Caribbean that used its raw materials to produce manufactured goods for export.[39]

If the arrangements created after 1940 were introduced because they were good for research and Britain's reputation, the mechanisms by which they were good for development were not well elaborated. A comment on research into levulinic acid described the anticipated relationship between research and industrial development as being one where 'There can be no doubt ... that when projects reach such a stage as this and reasonable profits can be foreseen from their pursuit the sugar industry, either alone or in collaboration with other industries, will come forward to finance them'.[40] In other words, if Wiggins ensured that information was available on new products, then business would take over. The responsibility for decisions around establishing new industry did not lie with the scientific staff at ICTA, or the colonial government. There was only one case in which direct government support for new sugar-based industry was forthcoming: the sugar cane wax factory that received modest support through the Barbados Pioneer Industries Legislation of 1951. The provision of such encouragement was entirely at the discretion of individual British Caribbean governments, and neither the Colonial Office nor members of the BWISA appear to have lobbied for any support for new factories. There is no evidence that Wiggins ever met with members of the legislature in Trinidad to discuss the provision of incentives for sugar-based industry.

In its mode of operation, the STL in Trinidad conformed to the principles set out in the early 1940s by the new research committees at the Colonial Office. The laboratory operated autonomously from the Trinidadian legislature and communicated directly with the CPRC in London on its research programme. Research was under the control of the scientists who worked in the laboratory, who were free to pursue their investigations along their preferred lines. Once results were publicised, it was left to business to decide if it would exploit any new discoveries. What was absent was any formal dialogue with the government of Trinidad or any other British Caribbean legislature about the place of sugar research in the development plans that governments in the region conceived after 1945. This was no mere oversight but was the outcome of decisions that had been made during the early 1940s at the Colonial Office in which great importance was attached to the idea that the quality of scientific research in the colonies could only be raised if scientific investigators were able to operate independently from colonial administrations. This arrangement, in which autonomy and therefore professional status was guaranteed, was said to be the only way to engage a scientist of the calibre of Wiggins, who had come to Trinidad from the University of Birmingham. The key value that informed the work of the new STL and its place in colonial development was 'freedom': the need to protect both the autonomy of scientists and also the operation of a free market without excessive government interference.

The Colonial Microbiological Research Institute

The CPRC was responsible for the creation of a second laboratory in Trinidad named the Colonial Microbiological Research Institute (CMRI). This institute was established to deal with the problem of finding profitable uses for surplus sugar in fermentation and to aid colonial industrialisation through work in microbiology. The CMRI developed food yeast, surveyed the microbes in tropical soils for antibiotic effects, looked for a way to tackle Panama Disease in bananas and worked to improve the industrial processes of cocoa fermentation and rum distillation. The significance of the creation of the first microbiological research centre in the tropical British Empire was exhorted by Hankey in his speech at the opening of the CMRI on 5 July 1948, in articles in the *Trinidad Guardian* and also by Simonsen in a BBC broadcast in the spring of 1948. Three main points were made in the publicity surrounding the opening of the new laboratory. One was the vital role that microbiology had come to play in health, and reference was almost always made to the discovery of penicillin. Secondly, there was emphasis on the importance of microbiology to industry, especially the brewing industries, but also agricultural industries, since microbiology could both tackle pests and improve fermentation processes.[41] The third point was that research into microbiology was an area where Britain had some ground to make up. Simonsen commented on the radio, 'the recent discovery of the antibiotics has indicated only too clearly this gap in our educational system'.[42] Hankey told guests at the opening of the institute that 'we are striking out in a new line, in which

British scientific research has lagged behind the needs of our time' and the laboratory 'fills a gap in the national equipment'. Britain's capacity to undertake microbiological research was of significance to Hankey as he was also chair of the Biological Research Advisory Board that oversaw the new programme of microbiological work at Porton Down, the site of Britain's biological warfare research programme.

To an even greater extent than the Francis Watts Laboratory, the CMRI was actively fashioned as an important symbol of the progressive and modernising temper of British colonial policy after 1940. As well as being described as a unique venture for Britain, filling a gap in British science, the CMRI was also said to place Trinidad firmly 'on the scientific map'. This elevation of Trinidad to an international centre for science was possible as the CMRI was said to undertake fundamental research into general issues that went beyond the needs of Trinidad to those of the Colonial Empire more widely; it was research into universal phenomena and problems. If the CMRI signalled Britain's commitment to developing and modernising Trinidad then the island's nascent modernity rested on the ability of the laboratory's work to transcend its locality and participate in the global circulation of knowledge.

The idea of funding research into microbiology had arisen on a number of occasions before it was discussed at the 7th meeting of the CPRC in 1944. At this meeting, members of the CPRC were given a list of new factory enterprises in the colonies for manufacturing soap, margarine and food yeast. Food yeast was derived from a specially selected variety of yeast grown on molasses and then processed into a pale yellow powder that could be incorporated into various foodstuffs.[43] The British government had high hopes for food yeast in the Caribbean as it provided a source of protein rich in vitamin B that could be effective in tackling the nutritional problems of the region. A government-run Colonial Food Yeast Company operated in Jamaica based on a culture that originated at the DSIR's Chemical Research Laboratory at Teddington, through the work of the microbiologist Dr A. C. Thaysen.[44] Similar factories based on samples of the Teddington culture were expected for Barbados, British Guiana and Trinidad.

Thaysen responded positively to a proposal from the CPRC that he pursue his research into food yeast in a new Caribbean institute built using CDW funds. Between April and July 1945 Thaysen was sent out to visit the West Indies, the US and Canada to consider the best location for the laboratory and to contact American scientists working on fermentation and microbiological problems. Thaysen visited Peoria and saw the equipment that had been developed there for the large-scale production of penicillin. He noted in his report how the laboratory was using *aspergillus terreus* to make itaconic acid as a basis for colourless plastics, something that Wiggins and his team took up as a research problem. After Peoria, Thaysen went to St Louis and met with the directors of the brewing firm Anhauser-Busch. The company was at the forefront of explorations into food yeast in the US and expressed much interest in the Jamaica food yeast factory. Thaysen reported back to the CPRC that the work at Anhauser-Busch was less advanced than it was in Jamaica but noted the popularity of the yeast amongst workers at the brewing firm's plant.

> In a well organised domestic science laboratory, Messrs Anhauser-Busch are studying the incorporation of their yeast preparation into typical American dishes which are made available to the staff of the breweries. The writer was told that such dishes are in great demand.[45]

Thaysen also went to Rutgers University and met with Selman Waksman, who showed him a preparation of the antibiotic streptomycin derived from the actinomyces genus of soil bacteria.[46] Thaysen speculated on the possibilities of surveying tropical soils to find antibacterial agents that could be used to tackle economically important agricultural diseases such as Panama Disease that blighted bananas.

Thaysen's report on his visit to Trinidad is interesting in revealing his concern that the projected CMRI should be located in a prominent place. As Simonsen and Robinson had before him, Thaysen expressed misgivings about the ICTA and was not prepared to see his institute located in its grounds at St Augustine in central Trinidad. The agriculturalist A. J. Wakefield wrote to Sir Kenneth Blackburne, Secretary of the CDW Org, on Thaysen's rejection of ICTA as a suitable site for the CMRI, 'This causes me no surprise. In this connection you are aware of all that I have done to try to get the authorities to see how shockingly bad things are at the college.'[47] Thaysen's preference was for the CMRI to be in the capital, Port-of-Spain. Sir Rupert Briercliffe, Medical Advisor to CDW Org, reported to the Colonial Office after a meeting with Thaysen, 'he wished to have his laboratory sited centrally and conspicuously so as to show people that colonial research means something'.[48] Thaysen engaged the services of the Bristol-based firm of architects W. H. Watkins and Partners, who had offices in Trinidad, and commissioned them to design his laboratory in the St Clair district of Port-of-Spain. The modernist building they designed was subsequently featured in *The Builder* magazine. The laboratory was formally opened in 1948 but Thaysen arrived in Trinidad in January 1947 with his family and assistant, Muriel Morris, to begin work in temporary accommodation.

The CPRC reported that Trinidad had been chosen as the location for a new laboratory that would investigate microbiology and fermentation in a tropical environment as it provided a good supply of raw materials in the form of sugar and it was relatively close to Peoria, where similar work was done, and so exchange of findings with scientists in the US could occur. There was some debate, however, as to the extent to which the government of Trinidad would have oversight of the laboratory. Simonsen informed Thaysen in 1947 that whilst a grant from the Research Fund of the 1945 CDW Act would fund the CMRI, it might be sensible to devolve administration of the money to the Trinidadian government so they paid the salaries of Thaysen and his staff. The advantage of this arrangement was that the workers of the CMRI would pay local Trinidadian rates of income tax rather than UK rates. In this arrangement a member of the Trinidadian legislature would be nominated to attend meetings of the CMRI.

The suggestion that administration of the CMRI might be done by the Trinidadian government prompted a furious response from Thaysen, who described it as 'a grave mistake that no amount of immediate saving could compensate for'.[49] Thaysen continued in his response to Simonsen:

> in your address to the Chemical section of the British Association, you referred to the plans for the future of our institute and very aptly pointed out that 'it is intended that this research institute should be a centre for microbiological research not only for the Colonies but for the Commonwealth as a whole'.[50]

Thaysen argued that this greater international standing of the CMRI would be compromised if the laboratory was attached to the Trinidadian government. So far, he said,

> I have been recognised as an outsider, working, not under instructions from the local authorities, but on behalf of the Colonial Office which in starting our institute has had the interest at heart, not only of Trinidad or the West Indies, but the whole Tropical Empire.

Aside from compromising the imperial standing of the CMRI by bringing it under local control, Thaysen also had little confidence in the competence of the Trinidad authorities, stating that administration by the government might mean the laboratory became the victim of 'inertia in official quarters', including slow or absent communications. It seems likely that Thaysen feared his freedom of action in setting out the research agenda of the CMRI would be compromised if the Trinidadian government was allowed to have a say. As with the Francis Watts Laboratory headed by Wiggins, the scientific independence of Thaysen was something that the Colonial Office believed it was important to protect. Charles Carstairs, Head of the Research Department of the Colonial Office, advised that it was wise to keep the CMRI 'a centrally administered show'. He made a comparison with the East African Agricultural Research Institute at Amani in Tanganyika, one of Britain's oldest research institutions. Carstairs claimed that Amani had been unable to achieve its intended stature as a regional centre for agricultural research after it became attached to the Tanganyikan government.[51]

After 1940, officials were particularly concerned with the status and meanings afforded to research work in the colonies. New, high-profile laboratories, staffed by world-leading investigators, were intended to be evidence of Britain's commitment to solving colonial problems, a rebuke to those who might criticise Britain's colonial record. Questions were raised, however, about how laboratories created along the lines of the CMRI might be perceived by the inhabitants of Britain's colonies. Harold Tempany, the Colonial Office's Agricultural Advisor, worried whether 'the public in the colonies themselves would not feel that same sense of ownership and responsibility for what was going on, even though the quality of research was somewhat higher'.[52] The utility and relevance of research work for the colony that hosted a new institution was considered less important, however, than the creation of stature. This demonstrates the extent of official concern with the public perception of British government actions after 1945, confirming the point made by historians that the CDW Acts were considered to have an important propaganda purpose in improving Britain's reputation as a colonial power. Highly visible British interventions to modernise and improve economic conditions were particularly necessary in the British West Indies, which had been the site of riots in the 1930s and a region where British imperialism was most open to scrutiny by America. Aside from a desire to see new laboratories sited in a prominent location, concern with image is also apparent in a comment made by Gerald Clauson of the Colonial Office at a meeting of the CPRC in 1946. Each year, the CPRC provided an annual report of its activities, published by HMSO and available for purchase. Clauson was recorded as asking for the forthcoming CPRC report to be rewritten: 'Sir Gerald Clauson was of the opinion that the language in which the report was drafted was that of a well-educated English man and would not be intelligible to the average colonial reader; he asked that it might be rewritten in a simpler style.'[53] One function of the annual report, then, was to promote the value and achievements of scientific research to readers in the colonies themselves.

Thaysen and the CPRC worked hard to shape the image of the CMRI through the Trinidadian press. The institute was the subject of numerous articles and reports during the 1940s and 1950s in Trinidad's daily newspaper, the *Trinidad Guardian*. In the publicity surrounding the formation and opening of the CMRI, the international standing of the institute was emphasised over and over again. On his arrival in 1947, Thaysen informed the *Trinidad Guardian* that the CMRI was not purely a Trinidadian institute but would be serving the empire.[54] In 1948 Simonsen informed the newspaper that no individual colonial government was paying for the institute, but it 'was a completely British Government venture' as a way of signalling that the reach of the laboratory went beyond Trinidad.[55] In one of a number of articles to mark the opening of the CMRI titled 'Micro institute to be world centre', readers were told how the CMRI would 'put Trinidad on the scientific map' and Thaysen was quoted as saying, 'No effort has been spared, in fact, to make the Institute worthy of British science and of its position as a world research centre.' The writer of the article then followed this point with the declaration

> The Trinidad Institute is we believe the only one of its kind in the British Commonwealth and one of the few such institutions to be found anywhere in the world. Its importance is attested by the number of distinguished scientists and other guests who have assembled for the opening and this Colony is fortunate to have been chosen as its home.[56]

The message was clear – the function of the laboratory was to tackle microbiological problems of international importance and the institute was therefore of great prestige. The benefits accruing to Trinidad might not be the direct ones, of a laboratory that undertook work intimately tied to the development of the island, instead they would come from being host to a world centre for scientific research.

Apart from the source of its funding, evidence that the CMRI was a world centre for tropical microbiology was supposedly found in the eminent scientists that came to work at the laboratory from overseas. Thaysen was described in the *Trinidad Guardian* as a 'scientist of the top rank', and the arrival of new colleagues W. C. Forsyth from the Macaulay Soil Research Institute in Aberdeen and J. E. Rombouts from the Netherlands both received coverage in the Trinidadian press. Whilst describing such individuals as 'top-grade staff', the newspaper also raised the question of whether there would room for West Indian staff at the institute on the basis that 'the chance to share in work of such an important nature should not be missed by talented and suitably equipped young men and women in the Caribbean'.

Thaysen's response to the enquiry about employing local staff was, 'you must remember that the institute to be formed is not purely a Trinidadian, but an Imperial affair. West Indians will be as welcome as anyone else.' In other words, there might be room for Caribbean staff but since the CMRI was a world centre, scientists from the West Indies would not be prioritised over workers from elsewhere.[57] Yet again, the CMRI was not primarily engaged with the development of Trinidad or other British Caribbean colonies. The focus was on the symbolic value of the research institution as an emblem of Britain's modernising agenda.

While the CMRI was celebrated as a laboratory that aimed to undertake 'fundamental research in microbiology', it was also concerned with improving industrial processes.[58] As with the work of the STL, the public emphasis on fundamental research did not preclude more practical enquiries. The programme of work determined by Thaysen began with research into cocoa bean fermentation, the disposal of rum distillery waste and the identification of antibiotics from soil organisms. In a note of November 1951, Thaysen wrote that there had previously been investigations into the microflora of temperate regions but not those of tropical ones, except the pathogenic ones. Referring to streptomycin, Thaysen noted the need for the systematic study of tropical soil micro-organisms, and he discovered one organism early on that was found to produce an antifungal substance that the CMRI subsequently patented under the name of comirin.[59] In addition to this work, Thaysen continued his research into food yeast for which he attempted to secure funds from the Colonial Food Yeast Company Ltd. The CMRI was also chosen to be a tropical centre for the housing of a reference set of non-pathogenic bacterial cultures available to industry and academia, named the Hankey collection. Samples of the cultures held in the collection were sent out on request to researchers around the world. As with the STL, the early years of the work of the CMRI were optimistic ones. In his role as President of the Royal Society, Robert Robinson said of the institute that

> it is destined, not only to play an important part in the development of Trinidad and the Caribbean, but also to give an example to the rest of the world. It will surely make a significant contribution to science and we hope to also to human happiness and prosperity.[60]

Conclusion

One goal of the CPRC was to see research into tropical commodities lead to the creation of new industry in the colonies themselves. When it became clear that ICI and other British chemical firms were unlikely to set up new factories in the British Caribbean, the CPRC turned to sugar manufacturers. The CPRC believed that diversification into new products based on sugar was most likely to be encouraged through the formation of a research association of the sugar firms that operated in the British West Indies. They hoped, as the DSIR had hoped previously, that if state funds were used to encourage firms to collectively invest in scientific research, members of the BWISA would then most likely take up the production of new sugar-based products. The industrial development of the British West Indies would result from the discovery of industrial uses for sugar and their by-products 'on the spot' in laboratories in Trinidad. Since these laboratories would undertake fundamental research, of general application, state funds were not being used to further the interests of one business over another. The actions of the CPRC were intended to carefully negotiate the need for some form of government intervention in order to stimulate industrial development without giving excessive aid to one single firm or otherwise interfering in the freedom of action that should exist for businessmen.

The CPRC made one of the Colonial Office's most direct attempts to stimulate the industrialisation of the British West Indies after 1945. Generally, the Colonial Office offered advisors and guidelines and rejected central authorities to plan and coordinate industrial development, tariffs or a Caribbean development bank. The absence of conspicuous British government agencies or a comprehensive policy for industrial development may well have contributed to the invisibility of Britain's industrialisation approach, both at the time and subsequently. The two laboratories created in Trinidad as part of a strategy to support industrial development were intended to be highly visible demonstrations of Britain's commitment to colonial modernisation. The fact that the sugar laboratory and the CMRI were intended to transcend a Trinidadian identity may well have produced a problem in the long run, however. It is not clear if the people of Trinidad ever felt ownership of these bodies, and their subsequent invisibility in histories of the Caribbean may come from the fact that they were not perceived as working for the Caribbean at the time. It is possible that whilst choosing the route of state-funded research as a means of economic intervention in preference to tariffs or government corporations might have had the benefit of avoiding political controversy, it might have also meant that these institutions were placed at the margins of the concerns of Caribbean politicians and intellectuals and the wider public.

Notes

1. Bonneuil, "Development as experiment", pp. 258–281.
2. *Colonial Research, 1943–1944*, Cmd 6529.
3. TNA, CO 852/482/11.
4. Butler, *Industrialisation*.
5. Weir, *History of the Distillers Company*, p. 316.
6. TNA, CO 852/579/4.
7. TNA, CO 852/579/4.
8. TNA, CO 852/579/4.
9. TNA, CO 852/579/4.
10. TNA, CO 852/579/5.
11. Weir, *History of the Distillers Company*, pp. 324–325.
12. TNA, CO 852/579/5.
13. TNA, CO 852/579/6.
14. TNA, CO 899/1.
15. *Ibid.*
16. *Ibid.*
17. Another institution funded by the sugar industry carried out research into sugar cane breeding in Barbados. The BWI Central Sugar Cane Breeding Station was established in November 1932.
18. BWISA (Inc), *Reports on Research Work 1943* (Barbados: Advocate Co, Ltd, 1943).
19. TNA, CO 927/203/3.
20. TNA, CO 899/1.
21. I. Varcoe, "Co-operative research associations in British industry, 1918–1934", *Minerva* 19 (1981), 433–463.
22. TNA, CO 899/1.
23. TNA, CO899/1.
24. TNA, CO 899/1.
25. TNA, CO 927/13/9.
26. TNA, CO 899/1.
27. TNA, CO 899/2.
28. *Trinidad Guardian*, Saturday, 17 March 1951.
29. *Colonial Research, 1950–1951*, Cmd 8303.
30. Colonial Products Research Council, *8th Annual Report, Colonial Research, 1950–1951*, Cmd 8303.
31. S. Clarke, "Pure science with a practical aim: the meanings of fundamental research in Britain, circa 1916–1950", *Isis* 101 (2010), 285–311.
32. Library of the University of the West Indies, St Augustine Trinidad (UWE), "Annual Report to the Advisory Committee for the British West Indies Sugar Research Scheme 1951/52".
33. UWE, "Annual Report to the Advisory Committee for the British West Indies Sugar Research Scheme, 1954/1955".
34. Lewis papers, Princeton University, Box 35, Folder 9.
35. UWE, "British West Indies Sugar Research Scheme Progress Report, 1948–49".
36. UWE, "British West Indies Sugar Research Scheme Annual Report, 1952/1953".
37. UWE, "British West Indies Sugar Research Scheme Annual Report, 1952/1953".
38. UWE, "British West Indies Sugar Research Scheme Annual Report, 1952/1953".
39. W. Scott, *The Industrial Utilisation of Sugar Cane By-Products* (Kent House, Port-of-Spain, Trinidad: Caribbean Commission Central Secretariat, 1950).
40. UWE, "British West Indies Sugar Research Scheme Annual Report, 1954/1955".
41. *Ibid.*
42. *Ibid.*
43. *The Canberra Times*, "Food yeast as meat substitute", Monday, 23 April 1945, p. 2.
44. A. C. Thaysen, 'Food yeast: its nutritive value and its production from Empire sources', *Journal of the Royal Society of the Arts* 93 (1950), 353–364.

45 TNA, CO 899/1.
46 TNA, CO 899/1.
47 TNA, CO 1042/132.
48 *Ibid.*
49 *Ibid.*
50 *Ibid.*
51 *Ibid.*
52 TNA, CO 900/1.
53 TNA, CO 899/1.
54 *Trinidad Guardian*, "Microbiological institute to get top grade staff", 12 January 1947.
55 *Trinidad Guardian*, "Chemist from Scotland to Join 'Micro' Staff", 2 July 1948.
56 *Trinidad Guardian*, "Micro Institute will be world centre", 4 July 1948.
57 *Trinidad Guardian*, "Microbiological Institute to get top grade staff", 12 January 1947.
58 "The Colonial Microbiological Research Institute, Trinidad", *Nature* (25 September 1948), p. 162.
59 TNA, CO 927/202/7.
60 "The Colonial Microbiological Research Institute, Trinidad", *Nature* (25 September 1948) p. 162.

CHAPTER FIVE

An industrialisation programme for Trinidad

It is practically a cliché in discussions of the post-war Caribbean to state that the British government did nothing to foster the growth of secondary industry in the British West Indies after 1940, and even purposively frustrated development of this kind. The original fault is said to lie with the Moyne Commission since the Commission's 1945 report did not expound the need for any major initiatives to foster the growth of industry in the region.[1] In the standard story, a period of indifference by Britain to the development of new industry was only brought to an end by the intervention of W. Arthur Lewis when he published recommendations for economic diversification in 'The Industrialization of the British West Indies' in 1950.[2] We are told that the adoption of Lewis's ideas by the governments of the British Caribbean marked the first phase in the pursuit of industrial development over the long term. Some versions place emphasis on the importance of the Puerto Rican experience as a model for Lewis's programme.[3] In all accounts, the influence of Britain, or British advisors, on the industrialisation programmes that were created by the governments of the British Caribbean after 1940 is written out.

The story that has emerged, in all its permutations, does not stand up well to close investigation. This chapter will reconsider the dominant narrative by paying close attention to the chronology of events that led to colonies such as Trinidad creating their first pioneer industries legislation, and by evaluating the relative importance of a number of individuals and proposals to the eventual character of the Trinidad ordinances. The claim that Lewis's work was decisive in setting in motion the first programmes for industrial development undertaken in the British Caribbean is easily overturned.[4] Legislation to encourage industrialisation in Trinidad was in place before Lewis published his famous article. In addition, Trinidad's policies for industrial development between 1950 and 1956 omitted many of the recommendations made by Lewis. They also did not conform to the Puerto Rican example in most important respects. The Pioneer Industries Act passed in Trinidad in 1950 offered modest concessions to industry, including income tax exemption and import duty relief on plant and machinery, and Trinidad's legislative council also provided marketing funds for advertising the benefits of Trinidad to foreign investors. The legislation did not create a development corporation to provide capital to entrepreneurs or set up factories, making the Trinidad approach different from that of Jamaica, where an industrial development corporation was created in 1952. Overall, Trinidad's industrialisation strategy in the first half of the 1950s bore more resemblance to the concessions for new industry advocated by the Colonial Office in London than the strategies promoted by Lewis or employed in Puerto Rico. This is unsurprising when we consider that one of the authors of the legislation passed in 1950 was an economic advisor seconded to Trinidad by the Colonial Office. This British advisor was Arthur Shenfield, an advocate of minimal state intervention in economic affairs (in contrast to Lewis), and later in life, President of the Mont Pelerin Society. This chapter shows that despite the threat presented by the US to the authority of British colonial rule, the Colonial Office was successful in steering policy for industry along lines it saw as desirable until the 1956 elections that brought Eric Williams to power. This success was achieved not by direct instruction by London but through the judicious use of expert advisors who promoted the more liberal road to development favoured by the Colonial Office. Only with the election of Williams did Trinidad embrace a different model, devised by Lewis. Lewis in turn drew upon the Puerto Rican programme for inspiration.

Development planning in Trinidad, 1945–52

Officials in London had come to express the view after 1942 that a degree of industrial development in Britain's Caribbean territories was necessary to provide employment and raise the standard of living. Whilst the Colonial Office had clear preferences with regard to the way industrialisation should proceed in the British West Indies, responsibility for working out the details of policies for industrial development lay with colonial governments. In the spring of 1950, the Trinidadian government passed a Pioneer Industries Ordinance that had its origins in an economic survey initiated by the Governor of Trinidad, Sir John Shaw, in 1947. The Economics Committee that undertook the survey contained individuals who had been recently elected to the

Trinidad legislative council after the implementation of constitutional reforms. Trinidad had held its first elections with full franchise in 1946 and the result was a legislature with nine elected members and nine members nominated by Shaw. This election brought the Trinidadian trade union leader Albert Gomes to prominence. Gomes became a member of the Economics Committee responsible for the survey of economic conditions on the island and was one of the authors of the Pioneer Industries Legislation.

The period after the end of the Second World War was one in which a consensus had emerged amongst metropolitan and colonial officials, colonial publics and non-British members of the Caribbean Commission on the necessity of industrialisation. It was not at all certain, however, from the perspective of the Colonial Office, that Britain would be able to guide its colonies along a desirable path of policy. Caribbean politicians elected to new legislative councils could be quick to perceive British self-interest at work in Colonial Office recommendations, believing that Britain sought to frustrate the development of industry that might compete with British firms. Alongside this lack of trust, there was also the issue of advisors representing the Caribbean Commission who promoted an alternative model of industrial development modelled on the Puerto Rican experience. The fact that the Trinidadian legislature was offered more than one vision of the path to industrial development was a serious problem from the perspective of officials in London. In a time of increasingly autonomous legislatures, the British government could not merely instruct its colonies in the British West Indies to follow its recommendations.[5] In this context, expert advisors assumed great significance and it was by the provision of advice that Britain sought to encourage the British Caribbean governments to formulate policy along its preferred lines.

In September 1947 Shaw formed an Economics Committee to undertake a comprehensive survey of economic conditions in Trinidad and make recommendations for the future. The decision to undertake a full economic survey was largely a response to criticism from the Colonial Office concerning Trinidad's inability to produce a coherent programme for development. Trinidad was notified of a £1.2 million allocation from the CDW Act in December 1945 and was required to present the office with a ten-year plan for approval.[6] The colony was described by officials as 'extremely backward', however, when it came to its capacity to produce a workable plan.[7] When it was finally released, Trinidad's ten-year development plan included large funds for road building, the extension of sanitation works, the development of the airport and the provision, or extension, of a number of hospitals. Apart from a concern with infrastructure, there was little indication of any planning for industrial development.[8] In April 1947, the Secretary of State for the Colonies, Arthur Creech Jones, asked the Governor to modify the colony's development plan so that more of the projects were 'directly revenue producing', or in other words, projects more clearly intended to improve the economic position of Trinidad. Specifically, the Secretary of State asked that more attention be paid to encouraging manufacturing industry. Creech Jones recommended that Trinidad request an economic advisor to travel to the island from Britain.[9]

The Economics Committee created in 1947 included amongst its members the economist C. Y. Shepherd of ICTA and a number of members elected to the legislature the year before.[10] One beneficiary of the 1946 election was Albert Gomes who won his North Port-of-Spain seat as a candidate for the United Front (UF), beating the famous nationalist leader Uriah Butler, who had contested this seat rather than fight in the oilfield area of the south where his party, the British Empire Workers and Citizens Home Rule Party, had its strongest support. The UF had been formed in early 1946 through the consolidation of a number of left-wing organisations. In 1946, 47 per cent of the population of the colony were black, 35 per cent were East Indian and the remainder were described in the census of that year as mixed, white, Syrian, Chinese or other.[11] Gomes was of Portuguese descent and it has been said that his success amongst voters was due in part to the fact he did not belong to the larger black or Indian populations of Trinidad and was able to unite factions that might otherwise be opposed.[12] After his election to the legislative council, Gomes agreed to sit on the Executive Council and was invited to serve on the Economics Committee and the Finance Sub-Committee.[13]

The economic survey was completed in 1949 and made a number of recommendations to address the difficult conditions that Trinidad was facing, including some economic diversification. In the post-war period, Trinidad was almost completely reliant on two industries, sugar and oil, and of these, the sugar industry was in poor shape. Production of sugar had fallen dramatically during the war. In 1944 an inquiry chaired by F. C. C. Benham, the Economic Advisor of the CDW Org, had described Trinidad's sugar industry as 'heading towards extinction'. Wartime conditions were responsible on two counts – soaring food prices had led independent cane farmers to switch to producing food crops. In addition, there had been an exodus of workers from the sugar estates, attracted by the better wages helping construct the American bases.[14] The result was that total labour in the sugar estates during crop time dropped to 16,700 in 1943, from around 25,000 in 1940.[15] The colonial government attempted to rehabilitate the industry by providing a planting subsidy, grants for the estates and a guaranteed price for the sugar harvest. There was also a move to improve the very low wages paid to sugar labourers. In December 1944 the All Trinidad Sugar Estates and Factories Workers Trade Union came to an agreement with the Sugar Manufacturers Federation for a 15 per cent increase in the rate paid to field and factory workers, plus holidays with pay.[16] Despite these attempts to revive the sugar industry, workers who left for alternative employment did not always return once the American bases were complete. The industry had still not recovered in 1947 and Trinidad was failing to reach the quota allocated to it by the British government.[17] The 1947 Economics Committee stated that improved living standards for all agricultural workers were necessary to attract people back to the land. Agricultural workers did not just need better wages they also required opportunities for education, better housing, medical care, and decent food and clothing, or in other words, 'a share of the privileges and duties of citizenship'.[18] Intervention by the Labour Government in Britain led to the creation of a Sugar Industry Labour Welfare Fund in 1947 across the British Caribbean, financed from a cess on sugar exports. It was used to provide loans at 1 per cent interest to sugar workers so they could buy or build a house and leave the notorious barrack housing on the estates, some of which was unchanged since construction in the mid nineteenth century.[19]

While sugar was the biggest employer on the island, by far the most valuable Trinidadian export was oil, which by 1947 represented 76 per cent of the total value of Trinidad's exports.[20] The oil production facilities in the south of the island had been modernised during the war and plant had been constructed to produce high-grade aviation fuel for Britain.[21] Workers for the oil companies received better wages and had a higher status than those on the sugar estates and this industry attracted migrants to Trinidad from other smaller Caribbean islands. Nationalist leader Uriah Butler had come to Trinidad from Grenada as an oil worker.[22] Despite the relatively higher wages paid in this industry, living conditions for labourers could be extremely basic, and relationships between black workers and white managers of companies such as Trinidad Leaseholds could be very poor. The racism of managers recruited from South Africa for work in the oil industry had been one of the grievances cited to investigators into the causes of the 1937 riots.

In the period after 1945 there were many episodes of strikes and riots amongst workers in the sugar and oil industry, and these signs of poor industrial relations caused great concern to colonial officials contemplating the potential for industrial development. In January 1947 a state of emergency was declared in the south of Trinidad after a strike shut down most of the oil fields. After oil wells were set alight in the Guapo district, the Governor imposed a curfew and created an exclusion zone that prevented anyone coming within 100 yards of the oil wells, tanks and refineries in an attempt to prevent further acts of sabotage.[23] The state of emergency was then extended to the whole island and the police were given the freedom to arrest anyone who did not move on request or give their name and address. Police used tear gas and batons after a crowd of supporters led by Butler invaded the Red House, the building housing the Legislative and Executive Councils in Port-of-Spain. The police also raided the headquarters of Butler's union in the capital, arresting hundreds of people after fighting broke out and a policeman was shot.[24]

Further strikes followed in May involving around 1,400 workers on the sugar estates, and there were more acts of sabotage in the oil fields. The legislative council invited the British trade union official F. W. Dalley to Trinidad to investigate. In his report, Dalley emphasised the importance of collective bargaining as the means by which workers should peacefully resolve their grievances with their employers rather than resorting to violence. The island's main daily newspaper, the *Trinidad Guardian*, condemned the strikes and called for responsible behaviour by the trade unions, while acknowledging that everyone knew that the high cost of living was a factor contributing to unrest.[25] The official figures showed the cost of living index at 221 in 1947, from 100 in 1935.[26] The Trinidadian people had endured hardship during the war, with severe shortages of food as shipping was diverted, and a high level of inflation. After the war, the high cost of foodstuffs was exacerbated by the dollar shortage as Trinidad was dependent on imported food items from North America. Between 1939 and 1947, imports from the sterling area shrank and those from Canada and the United States grew so that by 1947, 70 per cent of imports came from the dollar countries, compared to 51 per cent in 1939. The New World Group economist Edwin Carrington said of Trinidad in this period, 'the things which we produced we did not consume and those which we consumed we did not produce'.[27] In an attempt to alleviate the privation affecting the colony, a Food Controller was appointed in 1942 to regulate food imports, encourage the domestic production of crops and organise food distribution. A Price Control Committee created a rationing system and introduced government subsidies and fixed prices for rice, flour and condensed milk. Price controls were still in place by 1947 on imported foodstuffs such as rice.[28]

As well as facing rising prices, many workers in Trinidad struggled to find sufficient employment. A letter to the Colonial Office relayed the opinion of one judge in Trinidad that 'Uriah Butler is trying to stage a political comeback by exploiting the dissatisfaction of intermittently employed people with their insufficiency of income.' The letter writer, the Barbadian lawyer and one-time Attorney General of Trinidad, C. W. W. Greenidge, continued, 'the harbour strike was due to the fact that there are now two to three times as many waterfront workers as are needed in Port-of-Spain and that, while the daily wage of 2 Dollars, 56 cents (10/8d) was good, labour was diluted and very few of the workers earn more than two or three days wages a week'.[29]

The 1949 Economic Survey recommended that the production of citrus fruits, rice, cocoa and food crops should all be expanded so that the colony had a broader base of agricultural activity.[30] It also promoted the need for secondary industry to increase exports, reduce dollar expenditure and help to create purchasing power.[31] The war had not led to a large amount of new industry in Trinidad because of the difficulty in importing machinery, but some existing industries had expanded their capacity to meet increased demand for edible oils, margarine, soap, match-making and clothing. Simple engineering tasks had been taken up by local engineering establishments that carried out repair and replacement work for the sugar industry and for shipping.[32] In terms of government support for industrial development, Trinidad had a Local Industries Board in operation from 1941 with the function of exploring the possibility of establishing new industries. The achievements of this body were not readily apparent, however. In an article in May 1947 the *Trinidad Guardian* accused the government of inertia when it came to helping industry, citing the example of an aggrieved guava jelly manufacturer refused a sugar quota. The editorial asked, 'Where then is the official support for local industries?'. Trinidad, it stated, needed action not words.[33]

The main recommendation of the Industries Sub-Committee report was the creation of an Economic Advisory Board. The colony's new Economic Adviser, A. A. Shenfield, took up the position of chair of this board when he arrived in January 1949, two years after the Governor had agreed to the suggestion that such an advisor would be helpful.[34] Shenfield had a degree in economics from the University College of Wales and had studied law at the University of Birmingham. He joined the London and Cambridge Economic Service in 1937 and in the general election of 1945 he had stood as a Liberal MP although he subsequently broke with the party because of their endorsement of economic planning. The Economic Advisory Board that Shenfield headed, and which also included Gomes, made general recommendations to the Governor on economic matters and provided information to industrialists. It authorised the release of foreign exchange (usually dollars) to allow a firm to buy equipment or raw materials and gave special assistance to some producers of exports, including an allocation of sugar to producers of jam.[35] Shenfield interviewed applicants who sought government support to help start an industrial venture and he visited the premises of new firms.[36] Apart from this, the major achievement of the Economic Advisory Board was to draw up legislation that gave special concessions to new, or pioneer, industry.

The Income Tax (In Aid of Industry) Ordinance and an Aid to Pioneer Industries Ordinance were both passed into law in March 1950.[37] The basic provisions of the ordinances were an income tax holiday for new industry for five years, and duty-free imports of machinery and factory construction materials for the same period. Some special concessions were also provided for the cement and oil industries. Cement manufacturers benefited from a tax holiday and duty-free imports for ten years, and the oil refiner Trinidad Leaseholds was allowed to import crude oil for refining duty-free for twenty-five years from the 1 January 1949.[38] In addition, four areas were selected for development as industrial estates, with government providing roads and water.

Soon after Trinidad passed its new legislation, a complaint was passed to the Colonial Office from a Scottish beer producer. A Trinidad-based firm, the Caribbean Development Corporation, had established a brewery in Trinidad under the new Pioneer Industries Legislation and the business enjoyed exemption from income tax and a relatively low rate of duty. The managing director of the Scottish brewing firm, John Jeffrey and Co, complained to his MP that as a result of the advantages enjoyed by this Trinidad-based business, his firm had lost an order of more than 32,000 bottles of beer.[39] James Griffith, the Secretary of State for the Colonies, defended the Trinidad policy, stating that the measures to encourage new industry in Trinidad were 'perfectly legitimate and necessary'. Griffith pointed out in his response that the Colonial Office could not, in fact, intervene. Decisions about the operation of the Pioneer Industries Legislation lay with the Trinidadian legislature since 'as you will be aware, constitutional changes have recently been made which further devolve responsibility for local affairs on the Colonial Government'.[40]

Trinidad's industrialisation strategy, 1950–56

One historian has claimed that the initiatives put in place by Trinidad's Economics Committee meant that 'The Shaw Committee was clearly more favourably disposed toward industrialization than the Colonial Office.'[41] This conclusion is difficult to support given the expectation of the Colonial Office that industrial development would be part of the Trinidad's post-war plans and the fact that the office had sent an economic advisor from Britain to help with this. The authors of Trinidad's

Pioneer Industries Legislation were Shenfield and Gomes.[42] Gomes was re-elected to the legislative council in 1950 as a member of the Party of Political Progress, described by Ivar Oxaal as, 'a rather conservative, middle-class organization'.[43] In 1950, constitutional reform led to the creation of a ministerial system, and between 1950 and 1953, Gomes was Minister of Labour, Industry and Commerce and until 1956 he was effectively Chief Minister.[44]

A lack of sources makes it difficult to fully reconstruct the events and discussions that led to the creation of Trinidad's Pioneer Industries Legislation. Shenfield studied the provisions in Jamaica for the encouragement of industry that resulted in the Jamaica Pioneer Industries (Encouragement) Law of 1949 and visited Puerto Rico, before rejecting the work of PRIDCO as a model for Trinidad.[45] Something of Shenfield's views about what constituted appropriate methods for encouraging Caribbean development can be surmised from two articles he wrote for *New Commonwealth* in the 1950s.[46] Here Shenfield painted a rather downbeat picture of the economic position of Britain's Caribbean colonies. The situation, as he put it, was that they were all minor contributors to the world markets that existed for their primary products, unable to compete in terms of price with the high-volume, low-cost producers of oil, sugar, citrus fruits and rice. Referring to the growth of secondary industry in the British Caribbean, Shenfield said, 'Such development cannot take the place of the basic industries [sugar, bananas, oil], but it can make a valuable contribution; and it is sound in principle where, as in Trinidad but not in Jamaica, it is not founded on the protection of high-cost producers.' In both articles, Shenfield expressed his disapproval of the fact that Jamaica was using tariffs and quotas to help new industry. Shenfield noted that by 1958, Jamaica and Trinidad had experienced reasonable rates of economic growth but he disputed the claim that this was due to the growth of secondary industry in the case of Jamaica on the basis that its use of protection had meant a cost had been incurred by the colony's primary producers. Again, he expressed greater approval for the Trinidadian case: 'a fair number of new enterprises have been established, especially in Trinidad, not by tariffs or quotas but by taxation reliefs which have in fact involved little or no cost to the rest of the economy'.[47] Shenfield allowed that the encouragement of pioneer industries might require government support, stating his acceptance of the argument for 'infant industries' as a special case.[48] The exceptional measures that governments might contemplate to help new industries should not include protection by tariffs, however.

After his time in Trinidad, Shenfield took up a post as the Economic Director of the Federation of British Industries (FBI). Neil Rollings has shown the role played by Shenfield in disseminating neoliberal ideas amongst British businessmen during his tenure as Economic Director of the FBI between 1955 and 1967.[49] From the 1960s onwards, Shenfield had a successful career as an academic economist in the US. He was an early supporter of Hayek, between 1972 and 1974 was President of the Mont Pelerin Society, and was involved with the Institute of Economic Affairs and the Adam Smith Institute in Britain.[50] Shenfield's work in Trinidad indicates that at this point in his thinking he believed that there were acceptable modes of government support for new industries in places such as Trinidad, but his views on this were clearly different from contemporary economists such as Lewis and others, who advocated planning and a much larger role for the state in supporting industrialisation.

Gomes spoke of Shenfield in appreciative tones in his autobiography of 1974, describing him as someone with 'initiative, imagination and common sense' as well as being 'forthright and uninhibited' in expressing his views.[51] Speaking on the issue of industrial development in his autobiography, Gomes expressed views that echoed those of Shenfield in his *New Commonwealth* articles, criticising government methods that supported uneconomic industry and in doing so had adverse consequences for the agricultural sector.[52] The degree to which an accord in views existed between Gomes and Shenfield is also demonstrated by the critical comments Gomes made about trade unions in his autobiography, in which he quoted Shenfield in order to make his case. Gomes had begun his own political career as a trade unionist; he was involved in the formation of the Federated Workers Trade Union and was President-General from 1942 to 1944. He had been elected to the legislative council in 1946 as a representative of the UF, formed by the amalgamation of a number of organisations that represented labour. By the time he came to write his autobiography, however, Gomes was accusing workers of being as full of self-interest as their employers, saying that they obstructed development by their constant demands for greater wages. Gomes quoted an extract from a speech made by Shenfield at the Trinidad Chamber of Commerce in April 1964 in support of his point. Referring to Shenfield as 'the country's Economic Adviser fifteen years before, and mine for many years', Gomes endorsed the claim of Shenfield that the underlying cause of high unemployment in Trinidad was the behaviour of labour itself. Claiming that there was no need for the high level of unemployment that existed in Trinidad, Shenfield told his audience that the root cause was 'that the price of the service which the worker delivers to the employer, including the cost of stoppages and other interferences with production, is too high. Hence some of the labour cannot be sold and remains unemployed'.[53]

The need to avoid strikes and riots was a priority for the Trinidadian government as it came to implement its new industrialisation strategy in the 1950s. Both Gomes and the Governor Hubert Rance warned workers on numerous occasions that since industrialisation was dependent on enticing businessmen to invest in the island, Trinidad had to present an image that it was friendly to business.[54] In the Governor's address of 1950, Sir Hubert Rance told his audience:

> Customs and income tax concessions may attract and catch the eye, but fair play, sound industrial relationships and a healthy constitutional Trade Unionism are an essential requirement for an industrialization programme. Anyone who damages or endangers the Colony's reputation in this respect, be he aligned with employer or worker will do the country grave disservice. He will literally be robbing the unemployed of their chances of work with all that means for them and their families.

Rance warned his audience that riots on the island garnered bad publicity for Trinidad at a time when many places sought to encourage investment by manufacturers: 'a reputation once lost in this present scramble for new industries is not easily regained'.[55]

Gomes' political trajectory from union leader to 'moderate right-wing politician', in the words of the Colonial Office, made him extremely unpopular amongst some of his original supporters in Trinidad and he received numerous death threats during his tenure on the Executive Council.[56] Gomes claimed that he was as unpopular with business leaders as he was with labour because of his insistence on the need to dismantle the system of price controls and bulk purchasing schemes that had been introduced to relieve wartime shortages. Gomes defended his actions in this area with a statement that Shenfield seems likely to have endorsed: 'the market mechanism works to everyone's advantage when it is not trammelled by restrictions and controls'.[57]

The basic principle that informed the industrialisation strategy created by Shenfield and Gomes was to provide conditions conducive to foreign investment. Good industrial relations, tax holidays and duty-free imports were key. It is impossible to assess the degree to which contact with Shenfield contributed to Gomes' belief in the centrality of private enterprise for the industrial development of Trinidad. Whatever the origins of his change of political priorities, the strategy that Gomes promoted for industrial development between 1950 and 1956 did not include government loans or a development corporation that might operate its own factories. In contrast, Jamaica created an industrial development corporation in 1952 to provide government finance to business, to purchase land and build factories, and participate in the management of an enterprise that received a loan.[58] In addition, the Jamaican government authorised protective tariffs against imports in an effort to support local business.

With its rejection of tariffs and the provision of tax holidays and duty-free imports, the legislation of Shenfield and Gomes was in keeping with the recommendations of the Colonial Office. The Trinidad ordinances were underpinned by an attachment to a more liberal political economy than the more far-reaching and state-directed strategies advocated by Lewis or the Caribbean Commission. One account of the emergence of the Pioneer Industries Legislation in Trinidad claims that Gomes acknowledged in his autobiography the influence of Lewis.[59] There is no such reference by Gomes, and the fact that Lewis did not publish his recommendations for the industrialisation of the British West Indies until three months after Trinidad had passed its Pioneer Industries Ordinance makes this impossible. Furthermore, the article published by Lewis in 1950 on industrial development in the British Caribbean included a critical evaluation of the initiatives that had been recently introduced in Trinidad and Jamaica.[60] Similar criticisms were made by C. J. Burgess of the Caribbean Commission in a report on incentives for industry across the Caribbean discussed at a Caribbean Commission conference in 1952. In addition, a group of leading British businessmen expressed strong reservations about the ordinances, albeit from a rather different political perspective. A major focus of debate about the most suitable route to industrialisation was the model provided by Puerto Rico's Operation Bootstrap. For individuals such as Lewis, who advocated state-directed economic development, the Puerto Rican experience was an important and persuasive example of how real change could be effected by government initiative. For some representatives of British business who believed in a limited role for the state and freedom of action for entrepreneurs, the Puerto Rican programme was a red herring, providing little of value to the British colonies because of the very different conditions that prevailed there.

Lewis and Burgess

Lewis was invited to act as a consultant on industrial development to the Caribbean Commission by Eric Williams in his role as Deputy Chairman of the Caribbean Research Council.[61] Williams arranged for Lewis to visit Puerto Rico in 1949 and then ensured the wide circulation of the two articles that Lewis subsequently wrote for the *Caribbean Economic Review*: 'Industrial development in Puerto Rico' published in 1949 and 'The Industrialization of the British West Indies' published in May 1950.[62] Williams described the latter to Lewis as 'one of the most magnificent monographs I have ever seen', and said that the Secretary General of the Caribbean Commission, Lawrence Cramer, believed it to be 'compulsory reading for all people in the West Indies'.[63] The Commission regarded Lewis's work as their best opportunity to determine the future direction of Caribbean industrial development.

Lewis's two articles formed the centrepiece of a conference on industrial development held by the Caribbean Commission in Puerto Rico in 1952 attended by representatives from the Colonial Office and Caribbean politicians such as Gomes, and Alexander Bustamante from Jamaica. Both Lewis's articles and other material promoted by the Caribbean Commission contained critical commentary on the Pioneer Industries Legislation. The Puerto Rican programme was invariably presented first in the papers circulated to delegates, with subsequent discussion noting the ways in which other territories deviated from this model in terms of their policies.[64] Clearly, at the Caribbean Commission, Puerto Rico's industrialisation programme was the exemplar to be followed by all the Caribbean territories.

In the discussions at the conference it was noted that Trinidad and Jamaica had achieved the most of all the British territories in developing plans to encourage industry. The similarities between the policies adopted in Trinidad and Jamaica and the initiatives of Puerto Rico included provision of relief from import duties on machinery and tax exemption for new business. Trinidad and Jamaica had also undertaken publicity work intended to entice investors. They had contacted industrialists by mail, made promotional visits to the US and UK, and from 1950 onwards, they produced pamphlets – *Opportunity for Industry* in the case of Trinidad, and *Invest In Jamaica* produced by the Jamaican government. This publicity work was on a much smaller scale than that of Puerto Rico, however. Puerto Rico had permanently staffed offices in New York, Chicago and Los Angeles to advertise the benefits of the island to American investors.[65]

Despite some similarities between the initiatives of Britain's Caribbean territories and those of Puerto Rico, the delegates of the conference also identified substantial differences. Development boards or corporations did not exist in the British Caribbean at the time of the meeting (although by the end of 1952 Jamaica had created an industrial development corporation). Puerto Rico was the only territory to make direct government investments in industry, with the creation of government factories to produce cement, clay, glass, paper, shoes and leather, and PRIDCO had built factories to be leased to industry. In addition, Puerto Rico was also the only Caribbean territory with a government development bank, and this was considered particularly significant. C. J. Burgess, Executive Secretary for Economics on the Caribbean Research Council, noted that as the number of places that aspired to industrialisation increased, there was intense competition for capital investment and it was not at all clear in his opinion where the finance would come for new industrial ventures in Trinidad and Jamaica.[66] In his articles, Lewis recommended the formation of an industrial development bank to act as a lender of last resort, and an industrial development corporation that would act for the whole of the British Caribbean.[67] Overlooked in most discussions of Lewis's vision of British Caribbean industrial development is the fact that his plan for West Indian industrialisation was inseparable from his belief in the necessity of a British West Indies federation.[68] Federation of all the British West Indies colonies had been under discussion since the Montego Bay Conference of 1947. The Colonial Office hoped to counter the influence sought by the US in the region through the formation of a British West Indies federation that would give a stronger voice for the British colonies in the Caribbean.[69] Lewis was an advocate of federation on economic grounds and he believed a customs union was crucial to create large markets for manufactured goods and to prevent competition between industries arising in different Caribbean territories.[70] The Colonial Office shared Lewis's concern that uncoordinated industrial development would result in colonies vying with each other for foreign investment. The need for internal free trade and a consistent regime of duties between the British West Indies and countries outside the union formed a key part of the rationale for West Indian Federation. In the short term, a Regional Economic Committee (REC) was created that met for the first time in May 1951 and discussed, amongst other things, whether a more uniform pioneer industries legislation or some coordination of industrial development was needed.[71] Despite the concerns that were raised, this coordination did not occur.

In a discussion of the function of industrial development corporations, Lewis noted that Trinidad had earmarked four areas on the island as industrial estates – two near Port-of-Spain, one near Arima and one near San Fernando. Government intended to provide water and roads and offer leases to businesses for twenty-five years, with the option to renew.[72] Lewis was critical of this initiative, however, on the basis that setting aside land was not sufficient to attract businessmen; a government needed to provide factories.[73] Lewis also criticised tax holidays and duty-free imports for pioneer industries, saying that these types of concession were weak instruments to attract investment. The five-year income tax holiday, in particular, was said to be of little incentive to foreign industrialists if they still had to pay income tax to their home governments.[74] This criticism was borne out by the controversy that developed as it became apparent that income tax relief in Trinidad did not benefit British manufacturers as they still had to pay UK rates of tax.[75] The *Sunday Guardian* in Trinidad reported in October 1951 that not a single British company had established a factory on the island since the passing of the Pioneer Industries Ordinance in the spring of 1950. Gomes made the point in person to Colonial Office officials at a meeting in September 1951 that this tax anomaly meant that Trinidad could rely only on American and other non-British sources of capital.[76] Representations were made to the Colonial Office at the REC in 1951 in an effort to get Britain to amend the law to encourage UK firms to invest in the British Caribbean but the Treasury would not budge.[77] The *Sunday Guardian* also claimed at the end of 1951 that the UK was refusing to allow American and Canadian businesses in Trinidad to import plant and machinery from the dollar countries (a reference to the limited availability of dollars to the colonies for expenditure on imports) and questioned whether Britain was seriously committed to British West Indies industrial development or had 'thought it expedient to pay lip-service'.[78]

The outcome of the 1952 conference in Puerto Rico was recommendations for development banks and corporations derived from the reports of Burgess and Lewis. Burgess claimed that the existing policies for industrial development developed by Trinidad and other colonies were 'inadequate in scope and intensity', and governments were urged to assume far greater responsibility to 'generate a greater industrial momentum'.[79] For officials at the Colonial Office in London, however, the obstacle to industrial development was not the absence of government initiative and finance along the lines of Operation Bootstrap but rather the narrow range of industries that were ever likely to be economic. In a revealing remark that would seem to embody the Colonial Office attitude, Permanent Under-Secretary of State Hilton Poynton wrote to his colleagues, 'One would presume that if there were a lot of such industries, private enterprise would have got on to them.'[80] Poynton conceded, however, that the Colonial Office might further support Caribbean industrialisation by identifying those industries best suited to the region. The Colonial Office decided to consult eminent businessmen, and in 1952 a delegation of British industrialists was dispatched to the British West Indies to give their assessment of the strategies currently employed by the colonies and make recommendations for new industries for the future.[81]

The mission of British industrialists

The report produced by the mission of British industrialists who toured the Caribbean in 1952 was intended to provide support for British West Indies governments in improving their programmes of industrialisation and quite possibly to act as demonstration that Britain was not indifferent to the challenges involved in encouraging industrial development in the Caribbean colonies.[82] Its publication proved so controversial, however, that it threatened to further undermine the Colonial Office's claim that it was fully committed to the development of secondary industry.

After consultation with the Federation of British Industries, the Secretary of State for the Colonies, James Griffith, selected a delegation of UK industrialists comprising J. L. S. Steel (a director of ICI), L. Rose (L. Rose & Co, a manufacturer of lime cordial and lime marmalade), W. W. S. Robertson (W. H. A. Robertson & Co, engineers), Lt Col H. E. Pierce (Hall & Co) and G. H. Spencer of rubber goods maker George Spencer Ltd. The group undertook a tour of Jamaica, Trinidad, Barbados and British Guiana and visited over a hundred industrial establishments.

In their report, the authors admitted that several times during their tour, 'it was suggested to us that our findings might be influenced by a desire on the part of the United Kingdom manufacturers to restrain the development of industries in Colonies which might be competitive with exports from the United Kingdom'. The final report offered reassurance that the intention was to make recommendations that would first and foremost be of benefit to the Caribbean, even when they might negatively affect British exports. Alongside these claims, the report began with an upbeat appraisal of the prospects for future industrial development in the Caribbean, stating that 'the volume of industrial production will be nearly doubled within the next ten years, provided no unexpected and untoward political developments take place.'[83]

The positive appraisal of opportunities for industrial development was largely forgotten, however, in the response to the more critical aspects of the report. The mission of industrialists described their role to see if industry established in the British Caribbean was economically sound, and to assess the value of the strategies that had been adopted. The report condemned attempts to encourage industry that included the use of tariffs or quotas for imports.[84] Aside from protection, the report voiced strong criticism of references to the Puerto Rican experience and queried aspects of the operation of Jamaica's Industrial Development Corporation. The delegation expressed their approval of the loans provided by the corporation to business and the creation of industrial estates but not the fact that the Corporation was to run factories itself, commenting, 'throughout the world the history of industries started by Government or official corporations has not been a happy one'.[85]

AN INDUSTRIALISATION PROGRAMME FOR TRINIDAD

The repeated references by Caribbean politicians to the idea that the British West Indies should emulate Puerto Rico's Operation Bootstrap prompted the mission of industrialists to claim that Operation Bootstrap had in fact not been a complete success, and in particular that the four state-run factories had all made losses during the years they were government controlled. A more fundamental issue raised in the report was that it was the special relationship that Puerto Rico enjoyed with the United States that was in fact largely responsible for the success of the island's industrialisation programme. American manufacturers were able to take advantage of the low wages in Puerto Rico to set up factories and then import manufactured goods to the United States without incurring any customs duty since Puerto Rico was constitutionally part of the United States. The territories of the British West Indies did not enjoy a similar relationship with any country in the region that might provide a large market for their goods.[86] With regard to the possibility of UK manufacturers creating factories in the British West Indies to supply the markets of Central and South America, the authors of the report thought this was unlikely. Goods made in the Caribbean would cost more than those made in Britain because of the expense of importing raw materials, the freight costs for construction materials, and the absence of any favourable ratio of labour costs.[87] The growth of industry was therefore dependent on internal consumption.

In terms of the overall growth of industry in the British West Indies, the mission of industrialists stated that they had not found evidence to support the assertion that most industrialisation had occurred in the last few years, or that the rate of growth had been substantial in that period. They stated that the statistics available indicated that industrial production had been growing steadily for the past twenty years or so. This suggested that the introduction of pioneer industries ordinances from 1949 onwards had not made a significant difference. The report also made some specific criticisms of industries that had benefited from pioneer status but which appeared to merely undertake the final assembly of products made from parts imported from the dollar countries. It was suggested that two US companies, the Myerson Tooth Corporation, a manufacturer of false teeth, and the Simplex Time Recording Company, had created factories in Trinidad merely to take advantage of the access to the sterling area and were not intended to be long-term investments.[88] The factories did not have much capital equipment and did not employ large numbers of people:

> The Myerson factory is merely a packing station for consignments for the firm's customers in the sterling area. It is true that they polish up

the teeth after they have broken them out of the plastic sheets in which they are sent down from the USA; but that is a very small operation by a couple of girls.[89]

In the view of the mission, these factories would most likely close down once free convertibility with the dollar was restored.[90]

The delegation singled out two businesses based on the use of sugar or sugar cane as a raw material as potentially important, however. One was a factory in Jamaica that made anhydrous alcohol from molasses to be mixed with petrol for cars, and the other was a factory engaged in the production of sugar cane wax. A Barbados sugar cane wax pilot plant had been set up under the Pioneer Industries Legislation of that island. This project was the result of research undertaken with a grant from the CPRC and done partly in Wiggins' laboratory in Trinidad. The mission recommended market surveys were undertaken into the potential for sugar cane wax along with further technical work into producing a more consistent product.[91]

Despite the favourable comments on the future of industrial development in the British Caribbean, and the care that was taken to identity potentially successful ventures, the mission's report came in for heavy criticism in the British Caribbean, where it was taken as evidence that the UK wished to discourage manufacturing in the region in favour of its own exports. There was particular anger about the criticisms of the Pioneer Industries Legislation.[92] Albert Gomes told the Trinidadian press that the legislation was making a real contribution to industrial progress.[93] The Jamaican Minster of Trade and Industry complained that the report was not constructive. Officials at the CDW Org expressed the view that the positive aspects of the report could 'be swamped by the indignation at some of its brash and perfunctory generalisations'.[94] Hugh Foot, the Governor of Jamaica, wrote to the Colonial Office complaining about the report on 15 July 1953, saying he thought it would 'do more harm than good'.[95]

The negative response to the confidential circulation of the report caused the Colonial Office to debate whether they should distance themselves a little from the mission's conclusions, before eventually deciding that there was nothing to be gained from this since the remarks made about the methods and the scope for industrial development made by the mission were likely to be representative of the views of any potential investor who contemplated an enterprise in the British Caribbean, 'Their yardstick throughout was "will it pay?".'[96] The Colonial Office believed that the report fulfilled an important function in checking some of the grand claims that were being made about the scope for industrialisation in the British West

Indies.[97] Amongst politicians and the public in the British Caribbean the effect of the report was negative, however; it worked to further undermine the claim in London that the British government supported Caribbean economic diversification. It also did not succeed in its aim of prompting a re-evaluation of the path to industrial development that had been chosen by the British Caribbean territories, neither leading to reform of existing pioneer industries legislation, nor spurring changes to the role of Jamaica's Industrial Development Corporation. In fact, by the end of the 1950s, Trinidad came to adopt a new industrial development policy that moved the island's strategy closer to the recommendations of Lewis and the model provided by Puerto Rico. The turning point in Trinidad was the 1956 elections that saw Gomes lose his seat and Eric Williams' People's National Movement assume power.

The People's National Movement come to power

Williams' contract was not renewed by the Caribbean Commission and he left the organisation in 1955. After spending the rest of the year giving lectures in public in Trinidad and having discussions about the future of Trinidadian politics in private with friends such as George Padmore, C. L. R. James and Lewis, a constitution and manifesto was devised for a new political party to be launched in January 1956, the People's National Movement or PNM.[98] The PNM went on to win thirteen of the twenty-four seats on the legislature in the election of 1956, garnering 39 per cent of the vote overall. The Governor, Edward Betham Beetham, made arrangements that allowed the PNM to form the first government free of British control by giving two of the nominated seats to Williams' party and giving his guarantee that two of the *ex-officio* members would vote with the PNM.[99]

The PNM manifesto of 1956 was critical of the achievements of the Pioneer Industries Legislation passed in 1950 on the grounds that it had not made a notable contribution to increased employment or national output. Instead, Williams' party pledged to undertake a comprehensive economic survey and a study of industrial potential in order to create new jobs.[100] After the election, the new legislature passed a motion to consider the operation of the Pioneer Ordinances, and the chairman of the committee formed to do this, Patrick Hobson, visited Jamaica and Puerto Rico in July 1957 and studied Lewis's work on Puerto Rico's industrial programme.

The final report of Hobson's committee noted that at first glance it might seem that Trinidad's industrialisation programme had been a success. Fifty-seven manufacturers had been given pioneer status in the seven years since 1950, of which forty-three were still in operation. This was said to compare favourably with Puerto Rico, where the initial seven-year period of Operation Bootstrap between 1942 and 1949 had seen fifty-five industries created. The report continued, however, that despite this record, the achievements of Trinidad's industrialisation programme were 'pitifully inadequate'.[101] For a start, the process of endowing pioneer status was far too slow and firms were finding it difficult to get imported machinery and materials cleared while they waited for licences.[102] In addition, the number of people who had gained employment as a result of the Pioneer Industries Legislation was very small. The report made comparisons between Trinidad's initiatives to encourage manufacturing and those of Jamaica, noting that the amount of money used for advertising and marketing industrial opportunities in Jamaica was over ten times that spent by Trinidad.

Not unsurprisingly given Williams' promotion of Lewis's work when he was working for the Caribbean Commission, the solution to the issue of how to encourage greater industrial development in Trinidad was to be found in the recommendations of Lewis and the model provided by Puerto Rico.[103] The outcome was an industrial development corporation for Trinidad created in 1959. Amongst the functions of the corporation was the surveying of raw materials available to industry, recommendations for new industries suited to Trinidad and an expanded marketing campaign. The corporation also identified land for industrial estates, drew up plans for housing and schemes of civic improvement and issued business loans.[104]

The PNM won a second election in 1961 and Trinidad gained its independence in 1962. By 1963 there were ninety-nine pioneer industries in operation. The industrialisation-by-invitation approach meant that 80 per cent of the capital invested in industry was either American or British. Whilst the growth of manufacturing had been given a significant role in addressing the pressing issue of unemployment, it had very limited success. Only around 4,666 direct jobs had been created and yet rapid population growth meant the total labour force on the island had expanded between 1950 and 1960 by 100,000. In a longer analysis of Trinidad's economic performance between 1939 and 1967, the economist Edwin Carrington asserted that over this time, changes in the government or the introduction of new policies had not contributed to any major change in the performance and structure of the economy of Trinidad. In his words, the economy appeared to be largely unresponsive to the introduction of new initiatives and incentives.[105] His assessment of the impact of the pioneer industries legislation was that while the costs to government had been great, the scheme had failed to provide the gains that were originally forecast. The picture painted by Carrington in 1968 was one of foreign-owned companies providing little employment and with poor linkages to the local economy – 60 per cent of raw materials were imported, for example. The presence of foreign-owned companies had also failed to engender any entrepreneurial expertise amongst local businessmen. Despite the limited success of the Pioneer Industries programme, the Trinidadian economy grew at a rate of 8.5 per cent pa between 1951 and 1961.[106] This impressive growth was due to the performance of the oil industry that saw a seven-fold expansion in revenue between 1946 and 1956.[107]

In the longer term, none of the recommendations for policies to encourage industrial development that were promoted to Britain's Caribbean territories between 1940 and 1960 resulted in significance changes to the structure of their economies or reduced the high levels of unemployment. Dissatisfaction with the industrialisation-by-invitation approach led Caribbean economists such as Lloyd Best to produce a strong critique of Lewis's model. The differences between the ideas of Lewis, the Colonial Office and the US section of the Caribbean Commission that were so important in the 1940s and 1950s became irrelevant by the 1960s when Caribbean economists and politicians sought to reduce the dependence of places like Trinidad on foreign capital. The New World Group criticised Lewis's model for perpetuating the colonial legacy, and the PNM undertook a programme of nationalisation under the banner of 'Economic Independence' that by 1971 included government control of the sugar industry.

Conclusion

This chapter has provided a reassessment of accounts of the genesis of Trinidad's Pioneer Industries Legislation. The conventional story tells us that new initiatives for industrialisation in Trinidad were informed by the famous works of Lewis published in 1949 and 1950. This account has shown that Lewis's work was only important for the policies introduced to encourage industrial development in Trinidad after 1956, on the election of Eric Williams. A Colonial Office nominated advisor and advocate of laissez-faire economics, Arthur Shenfield, and the Trinidadian politician Albert Gomes created the first substantial legislation to encourage new industry in Trinidad in 1950. Their vision of development was more in line with the liberal approach favoured by the Colonial Office in Britain than the initiatives that were promoted by Lewis.

In the post-war Caribbean, the emergence of more than one vision of industrial development gave politicians in each of the Caribbean territories choices to make. One outcome of the divergence in views as to what constituted sound policy to encourage manufacturing was that Trinidad's first initiatives to encourage the process of industrialisation did not consist of the same techniques as other territories such as Jamaica. Caribbean political economy was not uniform in character in the post-war period, although it eventually converged on the model provided by the Puerto Rican experience. Interestingly, despite the differences that existed between the ideas promoted by the Colonial Office through their advisors and the model presented by Lewis, both sets of recommendations placed importance on a federation of British West Indies territories in order to increase the size of the market for manufactured goods and to prevent different colonies competing with each other in the industrial sector. The federation collapsed in 1962 and the question of how to secure an adequate market for goods produced in the British Caribbean was never resolved. The fact that the success of firms that participated in Operation Bootstrap often resulted from the free access that manufacturers on the island had to the large US market was largely ignored by those that sought to promote Puerto Rico as an appropriate model for Caribbean nations. This raises the question of the extent to which politicians and policy makers actually depended upon the Puerto Rico experience for a model of industrial development, despite their claims that this was so. The significance of this high-profile programme may have sometimes resided more in its utility as a rhetorical device for politicians who had a preference for state spending and state direction in order to produce much deeper and faster economic and social change in the British Caribbean than the Colonial Office in London believed was prudent.

Notes

1 Brereton, *History of Modern Trinidad*, p. 218; Pantin, *The Caribbean Economy*, p. xiv.
2 Kiely, *Politics of Labour*, pp. 5–6; datt Tewarie and Hosein, *Trade Investment*, pp. 2–3; Payne and Sutton, *Charting Caribbean Development*, pp. 2–3; Bernal, "The Great Depression", pp. 33–64. Mark Figueroa notes that the Colonial Office did see industrialisation as desirable and that Lewis was not responsible for the first wave of incentives for industry, M. Figueroa, "The academic economist as public teacher: lessons from W. Arthur Lewis and the Caribbean Policy Discourse", *Social and Economic Studies* 58 (2009), 10–11.
3 Kiely, *Politics of Labour*, pp. 5–6; datt Tewarie and Hosein, *Trade Investment*. Terence Farrell states that Caribbean governments did not actually adhere to Lewis's recommendations. He supports the idea of British resistance to industrialisation, however. T. Farrell, "Arthur Lewis and the case for Caribbean industrialisation", *Social and Economic Studies* 29(4), 52–75.
4 datt Tewarie and Hosein, *Trade Investment*, pp. 2–3; Payne and Sutton, *Charting Caribbean Development*, pp. 2–3; W. H. Griffith, "Lewis and Caribbean industrialization: policy, theory and the new technology", *The Journal of Developing Areas*

25 (1991), 207–230. Perhaps the strongest assertion of the idea of British opposition to industrialisation is found in Bernal, "The Great Depression", where he claims a 'deliberate colonial policy of discouraging industrialization' (p. 43) and asserts a refusal to provide any funds for industrial development on the part of the British government (p. 44).
5 TNA, CO 852/1037/1.
6 TNA, CO 852/874/5.
7 TNA, CO 852/874/5.
8 Havinden and Meredith, *Colonialism and Development*, pp. 253–255.
9 TNA, CO 295/642/4.
10 National Archives of Trinidad and Tobago (NATT), Box 6, no. 4, "Report of the Economics Committee, 1949".
11 K. Meighoo, *Politics in a Half-Made Society: Trinidad and Tobago 1925–2001* (Kingston: Ian Randle, 2003), p. 17.
12 I. Oxaal, *Black Intellectuals Come to Power: The Rise of Creole Nationalism in Trinidad and Tobago* (Cambridge, Mass: Schenkman Publishing, 1968).
13 TNA, CO 295/644/2. The other elected Trinidadian politicians who served on the committee were Errol Dos Santos, Victor Bryan, Chanka Maharaja and Gerald Wright.
14 NATT, "Report of the Committee Appointed to Enquire into the Sugar Industry. 1944", p. 5.
15 NATT, "Report of the Committee Appointed to Enquire into the Sugar Industry, 1944", p. 13.
16 NATT, Trinidad and Tobago, Council Papers 1945, "Industrial Adviser's Administration Report for the Year 1944".
17 Havinden and Meredith, *Colonialism and Development*, p. 208.
18 NATT, Box 6, no. 4, "Report of the Economics Committee, 1949", p. 11.
19 NATT, "Report of the Board of Inquiry into the causes and circumstances of a dispute in the sugar industry of Trinidad, 1955".
20 NATT, Box 6, no. 4, "Report of the Economics Committee, 1949".
21 Brereton, *History of Modern Trinidad*, p. 215; D. Edgerton, *Britain's War Machine: Weapons, Resources and Experts in the Second World War* (Oxford: Oxford University Press, 2011), pp. 181–187.
22 Brereton, *History of Modern Trinidad*, p. 180.
23 *Trinidad Guardian*, "Curfew imposed in St Patrick", 19 January 1947.
24 *Trinidad Guardian*, "Governor declares emergency", 22 January 1947.
25 *Trinidad Guardian*, Editorial "Trade unionist here to make a survey", 6 April 1947.
26 Bolland, *The Politics of Labour*, p. 526.
27 E. Carrington, "The post-war political economy of Trinidad and Tobago", *New World Quarterly* 4 (1967), p. 123.
28 Havinden and Meredith, *Colonialism and Development*, p. 209. Brereton, *History of Modern Trinidad*, pp. 213–214.
29 TNA, CO 295/642/4.
30 NATT, Box 6, no. 4, "Report of the Economics Committee, 1949", p. 3.
31 NATT, Box 6, no. 4, "Report of the Economics Committee, 1949".
32 NATT, Trinidad and Tobago, Council Papers 1945, "Industrial Adviser's Administration Report for the Year 1944".
33 *Trinidad Guardian*, "Local industries demand active official help", 6 May 1947.
34 TNA, CO 852/874/5.
35 TNA, CO 1031/74.
36 NATT, Box 6, no. 24, "Report of the Economics Committee, 1949".
37 NATT, Box 6, no. 24, "Report of the Economics Committee, 1949". This legislation was subsequently amended in July 1950, April 1952 and twice in June 1956.
38 TNA, CO 1031/74.
39 TNA, CO 295/655/7.
40 TNA, CO 295/655/.
41 Farrell, "Arthur Lewis and the case for Caribbean industrialisation", p. 56.

42 A. Gomes, *Through a Maze of Colour* (Trinidad: Key Caribbean Publications, 1974), p. 123.
43 Oxaal, *Black Intellectuals Come to Power*.
44 Ryan, *Race and Nationalism in Trinidad and Tobago*.
45 TNA, CO 295/655/7.
46 A. A. Shenfield, "Economic outlook, an upward trend", *New Commonwealth*, Special Caribbean Supplement 30 (1955), p. xi.
47 A. A. Shenfield, "Economic advance in the West Indies", *New Commonwealth* 35 (1958), p. 357.
48 Shenfield, "Economic advance in the West Indies", p. 357.
49 N. Rollings, "Cracks in the post-war Keynesian settlement? The role of organized business in Britain in the rise of neoliberalism before Margaret Thatcher", *Twentieth Century British History* 24 (2013), 637–659.
50 Rollings, "Cracks in the post-war", pp. 645–646.
51 Gomes, *Through a Maze of Colour*, p. 123.
52 Gomes, *Through a Maze of Colour*, p. 125.
53 Gomes, *Through a Maze of Colour*, p. 128.
54 NATT, Address by Rance, 24/10/52.
55 NATT Council Papers 1949, Governor's Address, First session of the new legislative council, Friday, 20 October 1950.
56 TNA, CO 1031/962.
57 Gomes, *Through a Maze of Colour*, p. 134.
58 Lewis Papers, Princeton University, "The Promotion of Industrial Development in the Caribbean, Report of Industrial Development Conference held in Puerto Rico February 11- 19 1952, Caribbean Commission, Kent House, Port-of-Spain".
59 K. I. Boodhoo, *Eric Williams: The Man and the Leader* (Maryland: University Press of America, 1986), p. 57.
60 Lewis, "The industrialization of the British West Indies", para 147.
61 EWMC, Box 063, E. Williams, "My relations with the Caribbean Commission, 1943–1955", Lecture Woodford Square, Trinidad, 21 June 1955, p. 27.
62 Lewis Papers, Princeton University, Box 14, Folder 1, letter Williams to Lewis, 13 August 1949; letter Williams to Lewis, 27 March 1950.
63 Lewis Papers, Princeton University, Box 14, Folder 1, letter Williams to Lewis, 27 March 1950.
64 EWMC, Box 044, letter Burgess to Williams, 10 May 1951.
65 Lewis Papers, Princeton University, "The Promotion of Industrial Development in the Caribbean, Report of Industrial Development Conference held in Puerto Rico February 11–19 1952, Caribbean Commission, Kent House, Port-of-Spain".
66 EWMC, 070 "Industrial Development Conference, Puerto Rico".
67 Lewis, "The industrialization of the British West Indies", p. 61.
68 Lewis, "The industrialization of the British West Indies", p. 61.
69 S. R. Ashton and D. Killingray, *The West Indies*, British Documents on the End of Empire, Series B, 6 (London: HMSO, 1999), p. xliv.
70 Lewis, "The industrialization of the British West Indies", p. 49.
71 TNA, CO 1042/149.
72 Lewis Papers, Princeton University, "The Promotion of Industrial Development in the Caribbean".
73 Lewis, "The industrialization of the British West Indies", pp. 62–63.
74 Lewis, "The industrialization of the British West Indies", p. 66.
75 TNA, CO 295/653/6.
76 TNA, CO 295/653/6.
77 TNA, CO 1031/851.
78 TNA, CO 295/653/6.
79 Lewis Papers, Princeton University, "The Promotion of Industrial Development".
80 TNA, CO 1042/149.
81 TNA, CO 1042/149.
82 TNA, CO 1031/75.

AN INDUSTRIALISATION PROGRAMME FOR TRINIDAD

83 TNA, CO 1031/75.
84 *Ibid.*
85 *Ibid.*
86 *Ibid.*
87 *Ibid.*
88 TNA, CO 1031/74.
89 TNA, CO 1031/74.
90 TNA, CO 1031/75.
91 TNA, CO 1031/75.
92 TNA, CO 1031/75.
93 TNA, CO 1031/75.
94 TNA, CO 1031/75.
95 TNA, CO 1031/75.
96 NA CO 1031/75.
97 NA CO 1031/75.
98 Brereton, *History of Modern Trinidad*, pp. 233–234.
99 Brereton, *History of Modern Trinidad*, p. 237.
100 EWMC, Box 547, "History of the PMN".
101 NATT, "First Interim Report of the Committee appointed to survey the pioneer industries programme 1950 to 1956".
102 *Ibid.*
103 EWMC, Box 757, "Untitled". "The State of the Nation, An Address Delivered to the Second Annual Convention of PNM, 28 September 1957".
104 NATT, "First Interim Report of the Committee appointed to survey the pioneer industries programme 1950 to 1956".
105 Carrington, "The post-war political economy of Trinidad and Tobago".
106 Brereton, *History of Modern Trinidad*, pp. 219–220.
107 Carrington, "The post-war political economy of Trinidad and Tobago", p. 141.

CHAPTER SIX

Bringing research 'down from the skies'

During the 1940s the scientists engaged by the Colonial Office were generally able to undertake projects of fundamental research in the chemistry of tropical products along lines of their own choosing. The notion that scientific researchers required the freedom to select their own research problems was a principle upheld by the CPRC and also officials at the Colonial Office concerned with the operation of the CDW Acts. By the early 1950s, however, officials at the Colonial Office were concerned that the work overseen by the CPRC was not making a tangible contribution to the economic development of the colonies. Officials complained that very few of the products developed through research were in commercial production. Colonial product research undertaken in Britain was subsequently reformulated with a focus on the analysis and assessment of tropical commodities in response to queries by business or governments. Most of the programmes of work previously done in university departments across Britain were terminated and investigation was instead concentrated under one roof in a new Colonial Products Laboratory. This marked the end of a period in which the emphasis had been on fundamental research in an academic setting and a return to the commercial intelligence work that had been traditionally undertaken by the Imperial Institute.

The work prosecuted in the two laboratories that had been created in Trinidad was not initially included in the reform of product research. By 1955, however, the programme of research at the STL was also being re-examined and there were concerns over the future of the CMRI. The CMRI and the STL had previously been promoted as institutions at the cutting edge of international scientific research whilst at the same time performing an important service in stimulating industry across the British Caribbean and wider Colonial Empire. The potential of new industry based on the use of cane sugar was endorsed in a report sponsored by the Caribbean Commission, and singled out for praise by the mission of British industrialists that had visited the Caribbean in 1952. When Colonial Office officials considered the achievements of Britain in terms of technical work of benefit to the colonies, carbohydrate chemistry was identified as an area where Britain could demonstrate it was a world leader. Wiggins' laboratory had attracted international acclaim and a number of organisations concerned with sugar research had sent funds and scientists to him, including the Sugar Foundation of New York. Despite all of this, none of the industrial ventures that emerged from the work in Trinidad were flourishing by the mid 1950s, and the Colonial Office had become increasingly critical of the work of the laboratory.

This chapter will consider the broader factors that limited the success of the CPRC programmes, such as changing political conditions in the colonies in the post-war period. Apart from this, there is a need to consider why, in a relatively short period of time, Colonial Office administrators seemed to have lost their enthusiasm for long-term programmes of fundamental research. The original vision of scientific research and colonial development did not place emphasis on rapid results, and the apparatus endorsed by officials for research in Britain and its colonies did not give a determining role to colonial governments or business in deciding what research would be done. In addition, the question of how the findings of research would be translated into practice was largely left unaddressed, following the model of the research councils at home. The early 1950s saw a significant change of heart at the Colonial Office and this chapter will consider both the external and internal factors that contributed to the demise of the agreement at the Colonial Office that undirected fundamental research had an important role to play in economic development.

Changes at the Colonial Office

The criticisms that were made about the work of the CPRC by the end of the 1940s represented a marked reversal of a previous approach in which officials had endorsed the claim that there needed to be a focus on programmes of fundamental research into the chemistry of colonial products. The consensus in the early 1940s between scientists and officials on the need to encourage work of this type emerged at a time when the Colonial Office faced a pressing need to restore the credibility of its policies. The crisis in the British Empire in the late 1930s in which Britain had to deal with widespread revolt in its colonies and critical scrutiny of its actions by other nations, most problematically the USA, had led Britain to announce in 1940 that a new era of colonial development was beginning. This included the declaration that Britain was committed to extensive fundamental research into tropical problems, a claim that was intended to be significant in a number of ways. The in-depth study of basic conditions was said to provide the foundation of knowledge upon which development plans would be based. In this way, fundamental research, as the study of fundamental issues, was assurance that Britain's new commitment to developing its colonies would work in practice. The Colonial Office were also concerned that attracting highly qualified and ambitious scientific researchers for work on colonial problems could be difficult as these individuals were most likely to seek university appointments in Britain. When officials announced that the Colonial Office would sponsor fundamental research they hoped that this would signal Colonial Office commitment to providing the sort of conditions for scientists that could be found through academic study or work with one of Britain's research councils.

When scientists from the research councils – the MRC, ARC and DSIR – advocated a commitment to fundamental research in the colonies, they sought most frequently freedom for scientists from oversight by individuals that were not qualified and experienced researchers themselves. The result was a string of research laboratories and stations in the Colonial Empire that enjoyed a significant degree of autonomy with respect to the administration of the colony in which they were based. This included arrangements that purposively limited the influence that the Agricultural, Medical or Veterinary Departments had over the research agenda of these new laboratories, on the grounds that the technical personnel of these departments were not well qualified in research. It is important to note, that in terms of the work that was done in these laboratories, and also some British universities, there was often a lot of short-term, practical problem solving. Elite British scientists did not intend to dictate that all work that went on in colonial research establishments had to be long-term fundamental research, their intention was to ensure that decisions about the work to be done were delegated to the scientists who worked in these institutions. Scientists were being given the freedom to pursue long-term and even speculative studies, *if they wished to*. The idea of fundamental research was deployed more for its rhetorical value than as a description of the work to be done; it denoted freedom for researchers to choose for themselves.

The consensus that was established in the early 1940s on the need for fundamental research was partly constructed from agreement between scientists and officials that colonial research needed to be endowed with a higher status. It was also forged, however, from two different sets of priorities. Scientists wished to see their preferred administrative arrangements introduced for research workers so that scientists controlled the research agenda. Officials found utility in the idea of 'fundamental research' for wider political reasons derived from the urgent need to demonstrate that Britain was taking its colonial responsibilities seriously and resist pressure from the US to place all dependent territories under the authority of some new international body in the post-war world. By the end of the 1940s, the most urgent economic and political issues facing Britain were different. There was far less reason for officials to support the notion that colonial development required programmes of fundamental research.

The suggestion that reform of research into tropical products was necessary was first raised when the work of the CPRC was amalgamated with that of the Colonial Products Advisory Bureau in 1947. The Colonial Products Advisory Bureau had previously been part of the Imperial Institute at South Kensington. It was transferred to the Colonial Office as the Imperial College of Science and Technology was scheduled to take over the building where the Imperial Institute had been housed since 1893.[1] In discussion about a merger between the Colonial Products Advisory Bureau and CPRC through some form of common supervision it became clear that the Colonial Office had concerns over the past work of the CPRC and saw the merger as an opportunity for reorganisation.[2] The result was a reduction in the amount of long-term exploratory investigation undertaken in Britain and a focus instead on the commercial evaluation of tropical commodities. This was less a new chapter in the history of work on colonial products and more the revival of the traditional functions of the Imperial Institute as an analytical and advisory service for colonial governments and British business.

The Colonial Office stated that its desire to reorganise the CPRC stemmed from concerns that the work undertaken in universities in Britain had not proven particularly useful. In the words of J. G. Hibbert, who had replaced Charles Carstairs as Head of the Research Department of the Colonial Office in 1947, product research needed to be 'brought down from the skies', and in the future, decisions over work needed to be made with 'its practical applicability to Colonial conditions as a primary consideration rather than possibly a secondary one'.[3] Hibbert was critical of the way that Simonsen had encouraged scientific researchers to pursue problems of scientific interest rather than directing attention to solving the problems facing the colonies, and he wanted more time to be spent addressing requests from colonial governments.[4] In private, officials expressed a low opinion of the abilities of Simonsen as an administrator and even as a scientist, and the chair of the CPRC, Lord Hankey, also came in for criticism on the basis that he had showed no inclination to rein in the scientists on the committee.[5]

A more critical attitude towards scientists involved in organising colonial research was apparent at the Colonial Office across the board from 1947. This coincided with the replacement of Charles Carstairs by J. G. Hibbert as Head of the Research Department. Carstairs had been very receptive to the claims made by the high-powered researchers engaged by the office about the necessity of programmes of fundamental research and he showed technocratic tendencies in his recommendations, including promoting the role of scientific data in planning. Hibbert was much more sceptical about the assertions of elite scientific advisors. He was a vocal participant in some fierce debates involving the scientists appointed to organise research in the colonies during the 1940s, most especially the Colonial Medical Research Committee (CMRC). This body had attempted to reorganise the structures that existed for the administration of research at the Colonial Office so that the CMRC had the final word on the allocation of scientists and funding in the colonies, relegating the administrative officials in the Colonial Office Research Department to a subordinate role.[6] Apart from concern that giving scientists complete authority resulted in projects that reflected the narrow interests of individual scientists rather than the needs of the colonies, there was also alarm over the fact that scientists in London did not seem to understand that their projects might be badly received by the subject people of the British Colonial Empire. Medical researchers associated with the MRC who sat on the CMRC seemed oblivious to the fact that the intrusive medical surveys they sought to fund might provoke unrest amongst colonised peoples, and they seemed ignorant of rising political consciousness in the colonies.[7]

The priorities and methods of the Colonial Office were reformulated during the tenure of Arthur Creech Jones, who became Secretary of State for the Colonies in 1946. Creech Jones expressed the aim of colonial policy in the Colonial Office's annual report of 1948 as being to 'guide the colonial territories to responsible self-government within the Commonwealth in conditions that ensure to the people concerned both a fair standard of living and freedom from oppression from any quarter'. The eventual independence of the colonies was set out as the goal of policy, although the Colonial Office believed that progress towards self-government would be gradual and require tutelage by Britain. New policies were created that were intended to increase

African representation on administrative bodies in the colonies in preparation for independence.[8] The new African Policy, largely created by the Head of the African Division, Andrew Cohen, had as its central aim the promotion of local government in the African territories as a way to offer Africans the experience of running of local services.[9] Officials believed recognising and working with nationalist sentiment would ensure the path to self-government was peaceful and orderly and newly independent nations would seek to be part of the Commonwealth. In the case of the British West Indies, constitutional changes had been occurring since 1945 that gave an increasingly greater voice to elected Caribbean politicians on colonial legislatures.

The raising of concerns by officials about the implications of centrally contrived research projects reflected the fact that greater opportunities were being given to the inhabitants of Britain's colonies to have a say in the running of their own affairs after 1947. Under Arthur Creech Jones, the Colonial Office moved away from its stance of the early 1940s, when it had promoted the need for direct intervention by the office in colonial affairs. This earlier approach had been prompted by frustration over the slow rate of progress in colonial development, attributed to the laissez-faire attitudes of the past and inadequacies of the colonial administrations when it came to planning development. Under this policy, Sydney Caine had celebrated the work of the CRC for taking the initiative when it came to organising scientific research. After 1947, a general trend towards devolved responsibility for policy meant that colonial governments could no longer be merely instructed by the Colonial Office with respect to new initiatives. Instead, progress was said to occur through a process of consultation and advice.[10] With this shift in the relationship between the Colonial Office and the colonies, the original arrangements for colonial research, in which research schemes in Britain's colonies were devised and implemented by committees in London, seemed at best to be out of step with developments in policy and, at worst, threatened to produce violent opposition from colonial peoples.

The notion that scientists did not understand the changing political conditions that existed in Britain's colonies was only one of the reasons why officials declared that the scientists they worked with could not be allowed to have complete freedom with regard to colonial research.[11] Hibbert believed that scientists did not generally make good administrators, as expressed in a confidential memorandum about the organisation of colonial research in 1947:

> I am afraid that it is true that scientists as a whole are frequently extremely temperamental and intolerant of any opinion opposed to their

own. This is occasionally due to conceit and the fact that they have achieved success in some specified and rather limited sphere. It is in other cases due to an inferiority complex. Scientists as a whole have not got a good business or organizing sense, although there are of course outstanding exceptions like Sir Henry Tizard. That is another reason why the final word should lie with people who have.[12]

When giving reasons why scientists could not be allowed complete authority over research, Hibbert and other officials invoked long-standing ideas about the differences between 'generalists' and 'specialists'. The archetypal civil servant administrator was said to be a generalist by virtue of a broad education in law or the humanities and traditionally a class background that meant a certain distance, and therefore impartiality, with respect to matters of trade and manufacturing. These endowments supposedly led to the broad and unbiased view of any matter necessary for sound and sensible policy-making.[13] Scientists, on the other hand, were specialists and therefore devoted to a single field. They were considered to be partisan, with a propensity to advance only those things they knew and favoured. In this memorandum of 1947, written in response to complaints that emerged from the recent African Governors' Conference about the low status of scientists, Hibbert seemed to be suggesting that the distinction between administrators and scientists was more than the product of education and experience but arose from differences in their psychological disposition. Whatever informed Hibbert's views, the important outcome of these claims about the abilities of scientists were summarised in the last sentence. By 1947, the Colonial Office asserted that when it came to scientific research in Britain's colonies, the 'final word' needed to lie with officials.

There is also the question of the impact of Britain's post-war economic crisis on official attitudes towards scientific advisors and colonial product research. In 1948, the colonies were being urged to increase the output of primary products and increase the speed of development in order to earn dollars and produce foodstuffs and raw materials in short supply in Britain. These goals were presented as compatible with the ambition of the CDW Acts and wider Colonial Office policy on the basis that an increase in production would strengthen colonial economies. A drive to increase primary products was therefore presented as part of the development of the potential of the colonies, something that was a necessary precursor to independence.[14] The revival of the claim that what was good for Britain was also good for the colonies belied the fact that controls on imports and currency meant the colonies were frustrated in their goal of securing capital goods and other materials for industrial development.[15]

At the Colonial Office there was a move towards ensuring 'the correct balance' existed between development schemes of social and economic value, meaning in practice a greater emphasis on economic schemes, particularly those that might contribute to increasing production of commodities, to the ultimate benefit of Britain. The Colonial Research Committee was renamed the Colonial Research Council in 1948 and was asked to maintain close contact with the Colonial Economic Development Council that had been created to ensure that the plans of the Colonial Office, and the ten-year plans being produced by colonial governments, had sufficient economic focus.[16] A Colonial Primary Products Committee was set up in May 1947 to consider how to increase outputs of a range of identified commodities, and it included amongst its members a representative from the CPRC. The need to identify sources of key materials from within the sterling area so as to reduce dollar expenditure had a direct impact on the research agenda of the CPRC when the committee was asked to help find a source of cortisone within the British Empire in 1949. Cortisone was an effective treatment for rheumatoid arthritis, but Britain's ability to purchase the drug from the US was threatened by the devaluation of the pound in 1949 and the rearmament drive prompted by the outbreak of the Korean War that placed further restrictions on dollar expenditure. The CPRC collaborated with the MRC and Glaxo in undertaking a search of plant steroids in the colonies that could be used to produce cortisone. This included engaging Thaysen to study ergosterol from yeast grown at the Food Yeast Factory in Jamaica, and the investigation of East African sisal as a potential raw material. Ergosterol did not prove to be as suitable as sisal. The latter furnished a source of hecogenin that could be used to produce cortisone, and Glaxo launched this product on the British market in 1955.[17]

While the CPRC took up a new project in direct response to Britain's post-war economic crisis, it was not the case that the council reorientated the overall direction of its work in 1947. While it seems likely that the new climate at the Colonial Office that privileged quicker and more discernible improvements in economic development was key to the criticisms that were being made about the past work of the CPRC by Hibbert and others, the committee was not immediately asked to refocus its work in a more practical or short-term direction. The annual report for 1947–48 produced by the council began by reiterating the claim that the work sponsored by the council 'must, of necessity, be of a long-term nature'. Readers were assured that this approach was appropriate because, the report claimed, the CPRC had received plenty of interest by firms in its research and was doing its utmost to ensure that results and patents were well advertised.[18] This did not amount to a significant turn towards ensuring greater, and faster, utility. It can be contrasted with the reformulation of the work of the Committee for Colonial Agricultural, Animal Health and Forestry Research that reported in 1948 that, 'the urgent need of the times is for a rapid expansion of production in the colonies'. The committee announced it had decided not to focus on programmes of fundamental research for the time being but to 'concentrate on problems of applied research for the solution of which the essential fundamental knowledge often already existed'.[19]

It seems likely that the Colonial Office decided to wait until Hankey and Simonsen stepped down from the CPRC before making any changes in the organisation of product research. In 1947 it was agreed that no encouragement should be given to either Simonsen or Hankey to stay on beyond their official retirement age.[20] The two men left in 1952 and the Colonial Office replaced the CPRC with the Colonial Products Council (CPC) on the 1 January 1954. Most of the original scientists that had comprised the CPRC retired or resigned, including Haworth and Heilbron. The chemist Sir Charles Dodds (Courtauld Professor of Biochemistry, University of London and Director of the Courtauld Institute of Biochemistry) replaced Hankey as the Chairman, and the new Director of Research was Dr R. A. E. Galley, who was also responsible for the research work of the Colonial Insecticides Committee.[21] The secretaries of the MRC, ARC and the DSIR remained on the body, along with Solly Zuckerman from the Lord President's Office. In a new initiative the Colonial Office appointed representatives from industry: Dr H. J. Channon from Unilever and Dr R. Holroyd from ICI, with the declared aim of ensuring that the future work of the council was of more obvious practical value.[22]

In the period between 1954 and 1959, the research work at British universities and research institutions was greatly reduced, with only the sugar research at Birmingham and forest products research at Princes Risborough continuing to be supported. The termination of the other research schemes in university departments dealing with derivatives of eugenol, citrus oils, and colonial fats and oils was done on the basis that very few products that were discovered or created by these projects, which in some cases had spanned ten years, were produced commercially. With respect to the sugar research at Birmingham University, the two compounds that had been considered most likely to have a commercial future were sodium levulinate and dextran. The first was found to be an effective anti-freeze, with several advantages over ethylene glycol. The CPRC stated in their report for 1947–48 that several firms had applied to work the patent for this compound, but it is not clear what became of this interest.[23] In 1955 it was reported to the CPRC that the firm Argus Chemical Corporation (New York) had offered to pay Wiggins' team, then based in the West Indies, for the information necessary to start commercial production of levulinic acid. The intention was to use this sugar derivative in the production of plastics, to be manufactured either on the Virgin Islands or in Trinidad.[24] This particular scheme was subsequently abandoned on commercial grounds.[25] The compound dextran, for which scientists at Birmingham developed a new synthetic process, fared better and was commercially produced as a blood plasma substitute.[26] After successful clinical trials in Britain and the USA, dextran was marketed under the name Intradex.[27]

Norman Haworth and Maurice Stacey had been working on a new process to make dextran from sugar since 1937 and their work came to fruition in 1943 when they developed a way to produce dextran from sucrose through the actions of a micro-organism named *Betacoccus Arabinosaceous* 'Birmingham'.[28] This work was supported by the CPRC after 1943, and in February 1949 the MRC put dextran through clinical trials at the Lister Institute, the Burns Research Unit at the Birmingham Accident Hospital and the MRC Blood Transfusion Research Unit.[29] The results of the trials were good; dextran resuscitated patients better than saline, it was free of contamination by disease, and it could be manufactured in large quantities and kept in refrigerated storage so could be stockpiled for use in an emergency. The CPRC promoted the discovery of a new process to make dextran from sugar for its contribution to the economic development of the British Empire, saying it was 'the first steps towards the establishment in the sugar-growing colonies of important industries based on sugar as a raw material, the ultimate goal which the Council had in mind when inaugurating its research programme on sugar'.[30] Hankey wrote to *The Times* in October 1950 in order to publicise the success of Haworth and Stacey and praise the CPRC for offering the prospect of developing industry in Britain's Caribbean colonies.[31]

Dextran was initially produced by the East Anglia Chemical Company in liaison with Haworth and Stacey, who had taken out a patent, and Haworth became a director of the firm from 1948.[32] The commercial production of the compound was not without its problems, however. The unforeseen issue affecting research into finding new uses for sugar in the period after the end of the war was the shortage of sugar in Britain. Any firm expressing an interest in producing a sugar derivative could find it very hard to secure sufficient quantities of the raw material to carry out pilot plant trials. The Ministry of Supply controlled sugar allocations amongst firms in Britain and maintained a system of sugar rationing for the British consumer until autumn 1953.[33]

The MRC wrote to the Ministry of Supply in 1947 asking that the East Anglia Chemical Company be given a quota of acetone (made from sugar) so it could manufacture a supply of dextran for clinical trials in Britain, stating that an alternative to blood plasma was badly needed because of the decline in blood donors in Britain since the end of the war.[34] The request was approved, and after the success of the MRC trials, commercial production at a factory in Aycliffe, Yorkshire, began in 1949 and orders of Intradex were placed by the Army, Air Force and Blood Transfusion Service.[35] The East Anglia Chemical Company changed its name to Dextran Ltd in 1950 and was then bought by Glaxo in 1952.[36]

Dextran was an exception to the more general trend in which the CPRC failed to find a firm willing to manufacture its products on an industrial scale. The commercial development of some of these was hampered by restrictions on the supply of raw materials during the 1940s.[37] Sugar and molasses had gone from being abundant, low-cost starting materials before the outbreak of the war to relatively difficult to acquire and expensive raw materials by the post-war period. To make matters worse, the government abolished the inconvenience allowance in 1945 that had previously ensured that alcohol fermented from molasses was an affordable starting compound for Britain's chemical manufacturing industry. By 1951, the prospect of any serious future for molasses and sugar as raw materials for chemical manufacturing in Britain seemed over when it was announced that ICI had switched to using oil as a starting material to manufacture synthetics. Chemical firms in Britain had lagged behind those of the US during the interwar period when it came to using components from oil to make their products. The Second World War was the turning point when Britain's need for aviation fuel led to substantial expansion in domestic oil-refining capacity. Further motivation for increases in capacity came after 1947 when the need to reduce dollar expenditure prompted Esso, Shell and British Petroleum to invest in new refineries, so that by 1954, Britain was the fourth largest refiner of oil in the world.[38] With expansion of oil refining came the increasing use of fractions from oil by chemical firms. ICI was a producer of aviation fuel itself during the war and entered into agreements with petroleum companies afterwards. The firm's move away from molasses was complete in 1952 when it opened an oil cracking plant at the enormous ICI site at Wilton on Teesside, using petroleum supplied by the Anglo-Iranian Oil Company.[39] Anglo-Iranian also formed the company British Petroleum Chemicals with DCL in 1947. DCL had been the biggest producer of alcohol for the chemical industry in Britain before the Second World War and it had been a major source of inspiration for the officials that created the CPRC as a way to resolve the crisis of the Caribbean sugar industry in 1943. After the Second World War, however, it no longer pursued a line of business based on alcohol from molasses.[40]

In September 1954, Charles Dodds, the chair of the newly reformulated CPC, reopened the discussions over the future direction which colonial products research should take. Dodds told Hilton Poynton of the Colonial Office that he was worried that 'the Colonial Office was not really getting value for money which it was spending on Colonial Products research'.[41] Echoing the earlier comments of Hibbert, Dodds expressed the view that research in the past had failed to address the problems of economic development. The decision was made to change the name of the Colonial Products Advisory Bureau to the Colonial Products Laboratory and make it the main location for research.[42] It was also decided that the CPC should be 'more rigorous in our scrutiny of some of the grants made to universities and other research workers for "farmed out" work'.[43] The advantage of this arrangement, in which research funds were no longer mainly spent on work outsourced to universities but done in the Colonial Products Laboratory was that it brought colonial products research under the direct control of the Colonial Office.[44] New premises were opened on Gray's Inn Road in December 1957 and the title of the laboratory was changed again to the Tropical Products Institute. The word 'colonial' was dropped in an attempt to encourage newly independent territories to continue using the services of the laboratory.[45]

In the period between the establishment of the CPC and the move to new premises, meetings of the council were dominated by discussion of the work of the new Colonial Products Laboratory, with very little of the plans and objectives of the original CPRC remaining. In general, the work of the laboratory consisted of responding to business and government enquiries that related to the whole range of natural products found in the Colonial Empire, and occasionally outside. Investigations were short term and applied in their approach, orientated towards the solution of specific queries, in contrast to some (but not all) of the more free-ranging and exploratory investigations sponsored by the CPRC. The Colonial Products Laboratory and then the Tropical Products Institute typically advised British business on the suitability of producing colonial agricultural products as commodities or the condition of markets for products. Technical advice was offered on processing techniques and an analytical service was provided for assessing the quality of colonial produce.[46] In April 1959, the Tropical Products Institute was transferred from the Colonial Office to the DSIR and the CPC replaced by the Tropical Products Institute Committee.

After 1953, the CPC largely provided a commercial intelligence service, much as the Imperial Institute had done in the past. The revival of the assessment of colonial products as the key function of product research, and a decline in the isolation of the constituents of tropical products and time-consuming exploration of their chemical pathways, reflected the changing political and economic conditions under which the Colonial Office operated. After 1947, new demands were placed upon the colonies because of domestic economic needs, and officials were asked to ensure that the initiatives they oversaw prioritised rapid economic development. In addition, the research committees at the Colonial Office found they were operating under different financial circumstances by the end of the 1940s. During the early years of the operation of the CDW Acts, the Colonial Office had struggled to spend its annual research allocation because of a shortage of scientific personnel and equipment. One of the big challenges that faced officials was ensuring that colonial research was attractive enough that highflying British scientists might consider a career in the British Colonial Empire. The idea that the Colonial Office had to compete with universities and research institutes run by the MRC, ARC and DSIR for scientific manpower meant that officials were easily persuaded that they needed to provide working conditions similar to those enjoyed by ambitious scientists at home. Officials were informed by the elite representatives of the research councils that advised them that this meant affording scientific researchers a great deal of latitude in choosing their research problems and making arrangements to ensure scientists were supervised only by other well-qualified scientists and not less-well-qualified technical staff or administrators. By the early 1950s, however, things had changed. In March 1949 the CMRC were told there was a sudden freeze on the spending of research funds as the annual limit of expenditure for 1949/1950 had been reached.[47] Again, in 1951 some research committees were told to rein in their spending as the projects they proposed outstripped the funds available to them.[48]

By the early 1950s there were also changes in the anticipated timescale of decolonisation. An emphasis on long-term projects of fundamental research was uncontroversial in 1943 when decolonisation was described as being at least a generation away in Britain's African colonies.[49] Official expectations about the likely timing of independence changed a great deal from 1947, making short-term results that had clear benefits much more desirable.[50] Taken together, the combination of a perception that the pace of change with regard to self-government had increased, Treasury demands for a focus on rapid economic change, and the fact that research fund finances were now tight, produced a climate in which officials asked whether they were getting 'value for money' when it came to research.⁵¹ A greater sense of urgency and increased accountability when it came to spending tended to work against an approach that privileged freedom for scientific researchers to pursue projects over the long term, above other considerations.

The factors that had underpinned the consensus over the need for an expansion of 'fundamental research' in the early 1940s did not pertain by the early 1950s. The passing of the 1940 CDW Act had occurred at a moment when Britain needed a powerful demonstration that it was taking its colonial responsibilities seriously. The goal was to offset revelations about the extent of deprivation in the British West Indies and elsewhere by an announcement that a new era of social improvement and economic progress had begun. The fact that the term 'fundamental research' was broad enough to encompass a range of meanings and inflections was important for building the consensus that existed before 1947. Officials embraced the idea of fundamental research for its promise of secure knowledge as a platform for development; the scientists that worked with officials used the idea of fundamental research first and foremost as a synonym for freedom from oversight by non-scientists. Officials were not unsympathetic to the promotion of scientific autonomy – they understood that this was an important component of the professional identity of scientific researchers and therefore acknowledgement of this was needed to attract new recruits – but the rhetorical value of a commitment to substantial expansion of fundamental research across the Colonial Empire lay elsewhere for them. By the end of the 1940s, the Colonial Office was dealing with a different set of priorities that were not so compatible with the idea of long-term, in-depth investigation of problems that had been selected by scientists without the input of business or colonial governments. The different economic and political conditions that emerged after 1947 tended to create an emphasis on faster and more tangible change whilst avoiding actions that might provoke unrest. Quick returns were needed that would increase the speed of development in the colonies and bring economic relief to Britain. These benefits needed to be provided under more straitened financial circumstances at the Colonial Office and had to fit in with the new and shorter timescale of decolonisation.

The laboratories in Trinidad

While the research programmes of the STL at the ICTA and the CMRI were not initially subject to the same scrutiny as the projects funded by the CPRC in British universities, by 1955, the CPC was beginning to express concerns over the utility of the work in Trinidad. New industries based on the production of food yeast, power alcohol, levulinic acid, sugar cane wax and ammoniated molasses as cattle feed were all failing to thrive in the West Indies, and by 1958, Wiggins' successor at the STL, Dr W. S. Wise, rejected the notion of further research into by-product processing. Instead, the laboratory shifted its attention to technical problems arising through the sugar-extraction process. By 1961, the STL had been closed and research into improving the process of sugar manufacturing was transferred to the Faculties of Agriculture and Engineering at the University of the West Indies. The CMRI closed in the same year.

Members of the CPC visited Wiggins' laboratory three times between 1954 and 1956 to report on the operation of the sugar technology scheme. The first visitor, Alexander Todd, was generally complimentary, reporting to the CPC the great potential of the sugar cane wax factory that had been given Pioneer Industry status in Barbados. The Barbados factory made a wax from locally grown canes that retailed at £500 per ton, comparing well with carnauba wax that sold at £900–£1,000 per ton and Cuban sugar cane wax at £600 per ton, both of which were used by the Johnson Co in the USA. Todd was also positive about levulinic acid, a plant for which was planned on the Virgin Islands by the Argus Chemical firm. He reported the news that the Quaker Oats Co had gone into partnership with the US chemical firm Dupont to produce another lucrative sugar derivative, furfural, in Puerto Rico as evidence of the viability of factories that used sugar to make useful chemical compounds.[52] Todd sounded the first note of caution, however, when he observed while there was evidence of interest in the research of the STL, it was not coming from the sugar manufacturers in the BWISA, the research association jointly run by sugar companies and the Colonial Office to fund Wiggins' laboratory.[53]

With regard to the CMRI, Todd praised the work that was being done, such as the study of the microbiological processes that occurred during fermentation of cocoa beans, but said he felt the small laboratory was too isolated. The CMRI was located in a suburb of Port-of-Spain, some distance from the concentration of scientists at the Imperial College of Tropical Agriculture in central Trinidad. Todd's comment was the beginning of lengthy but ultimately unresolved discussion about moving the CMRI to the ICTA, the University College of the West Indies in Jamaica or the Colonial Products Laboratory in London. The first director of the CMRI, A. C. Thaysen (who resigned in 1951 and was replaced by W. G. C. Forsyth), had rejected the suggestion that the laboratory be established at the ICTA on the grounds that he wanted the CMRI to be in a more prominent public location in order for it to be a highly visible symbol of scientific progress for Trinidad. Thaysen's communications show him to be an ardent defender of his own freedom when it came to determining the research programme of the CMRI and it is likely that he feared association with the ICTA might infringe this, something that was especially problematic as he did not appear to rate the college very highly. Visitors to the CMRI during the 1950s were concerned by the low staff morale, however, and believed the isolation of the unit was responsible. Todd was recorded as saying at a CPC meeting in 1956:

> He did not understand how it [the CMRI] had come to be sited in Trinidad where it was cut off by busy city communications from the ICTA which was the only place in the Island with the proper sort of research climate. The Institute was a small unit in a remote Island in a very large region.[54]

It seems that the efforts that had been made to elevate the CMRI to the status of a beacon of progress and international scientific advance, promoting Trinidad from merely being 'a remote island' and instead placing it on the 'scientific map', were not readily apparent to the next generation of scientific advisors at the Colonial Office. The CPC were concerned that the laboratory was not tenable in the long term and thought that transfer to another location was necessary if it was to survive after Trinidadian independence.[55]

The next CPC visitor to Trinidad was P. C. Spensley, who was a great deal more critical about the work of the STL than Todd had been. Spensley visited the STL in 1955 and commented that it had the atmosphere of a university department rather than an industrial research association laboratory of the type funded by the DSIR in Britain. He said that the laboratory was not paying enough attention to the needs of the sugar industry, it was not adequately exploring the commercial value of products, and not enough pilot plant work was being done. The CPC heard that at the meeting of the Sugar Research Scheme Advisory Committee held that year, the BWISA had requested that the laboratory provide more commercial information to them. They asked that Wiggins set out very clearly each phase from discovery of a chemical to pilot plant production to full-scale commercial factory production and provide an indication of the economic potential of a product by reference to market surveys and production costs.[56] In private, R. F. Innes, Research Director of the Sugar Manufacturers Association (Jamaica) Ltd, told Spensley that 'considerable elements of the BWI Sugar Industry are uneasy about the lack of concrete results so far achieved, and also feel that insufficient attention is being paid to the actual problems being experienced by producers'. He said that Wiggins did not visit the sugar factories enough.[57] Spensley commented that Wiggins was 'an organic chemist of academic inclination and ambition' and raised a query about whether it was most appropriate to locate the STL at the ICTA, a teaching establishment.[58]

Spensley also criticised the freedom that had been given to Wiggins, but on different grounds from those that had been previously invoked at the Colonial Office. Spensley complained that the firm that had applied to work the new process to produce levulinic acid developed at the STL was American and planned to set up its factory on the US Virgin Islands. Little benefit from the discovery of the new process to make levulinic acid was therefore accruing to either Britain or its colonies, apart from the royalties that would be paid out on Wiggins' patent. Spensley raised the question of whether allowing foreign firms to benefit from the research at the STL conformed to the original terms of the research association. He was also critical of the fact that Wiggins had accepted funds for research and fellowships from foreign organisations – the STL had been given a prestigious grant by the Sugar Foundation of New York and money for a fellowship by the Hawaiian Planters Association, something that had been celebrated as evidence of the standing of the laboratory at 'the forefront of the Sugar Research Institutes of the World'.[59] According to Spensley, developments of this type distracted the laboratory from its proper purpose of providing benefit to the BWISA.

Dodds and Galley, the Chair and Research Director of the CPC respectively, visited the Caribbean in May 1956 and attended a meeting of the BWISA. This was Wiggins' last meeting with the association as he had resigned his position and was due to leave Trinidad in September. In his subsequent report, Dodds echoed the criticisms that had been made by Spensley. Again the failures of the research at the STL to generate commercially successful products were attributed to Wiggins, who was said to have not adequately considered the economic implications of the laboratory's research. This time Dodds and Galley were able to refer to the recent closure of the sugar cane wax factory in Barbados that had finished operating after four years of experimentation and development. The reason given for the closure was that the factory in Barbados was 3,000 miles from the US market where there were already firms in position producing waxes.[60] Dodds used the end of this venture as an example of the 'the neglect of economic considerations that preceded some new projects in the West Indies'. In future, research schemes needed to be preceded by proper assessment of the potential market, possibly by using consultants, and there needed to be much more contact with colonial governments.[61] Criticisms were also made about the utility of the work of the CMRI. The laboratory had isolated a compound it named comirin from a local bacterium and promoted it as an anti-fungal agent. A patent was taken out by the

National Research and Development Corporation but then further research by the MRC and the Antibiotics Research Station cast doubt on its usefulness, and by 1956, the CPC decided to stop pursuing it.

One outcome of the criticisms that were made about the STL was that the CPC decided to reduce the money for the sugar scheme. This marked the end of a period in which sugar had been given priority over other areas of product research funded by the Colonial Office. Spensley wished the CPC to broaden its support for other industries and added that it was not desirable for the CPC to continue to fund a project in which it had no direct say.[62] The dissatisfaction expressed by the CPC about the work done at the STL, and to some extent at the CMRI, can be understood as part of the more general concerns that were being raised about the work that had previously been funded by the CPRC, most importantly that practical outputs had never been given sufficient priority. With the retirement and then death of Norman Haworth, and the replacement of the original chemists such as Simonsen, the STL and CMRI no longer had a cheerleader. The hopes of officials such as Caine for new industry in the Caribbean based on surplus sugar had been eroded during the 1940s as it became clear that chemical firms in Britain were not persuaded of the viability of this vision, even with Colonial Office expressions of support for schemes of this type. The claim that sugar could be a raw material for industry in Britain had far less credibility now that it was clear that British chemical manufacturing was investing in ethylene and other derivatives from fractions of oil. The only real prospect for sugar was that it could be used to create ventures in the Caribbean itself, possibly serving a regional market or as the raw material for exports to North or South America.

The hopes of Simonsen, Caine and Stockdale that the British West Indies sugar manufacturers would diversify their interests turned out to be misplaced. The annual reports of the BWISA between 1943 and 1955 indicate that manufacturers saw their future prosperity lying overwhelmingly with increases in the volume of sugar production, and the negotiation of a guaranteed price from the British government for sugar exports.[63] In May 1956, the BWISA told the CPC that they did not want another academic research worker, 'who would try to get them to enter the chemical industry'. Instead, they wanted a sugar technologist to head the STL and a focus on improvements to the technical process of sugar manufacturing. Asked what their aim was, the BWISA responded that they wanted to produce sugar more quickly and cheaply.[64]

There were a number of wider political and economic factors that worked to discourage sugar producers from diversifying into new chemical derivatives of sugar in the late colonial period. In April 1949 the

Labour Party announced its intention to nationalise sugar refining if it won the next election.[65] The Colonial Office debated whether the subsidiaries of Tate & Lyle, such as Caroni in Trinidad and the West India Sugar Company in Jamaica, would be nationalised.[66] The Ministry of Food told the Colonial Office that the Labour Party intended to take over any sugar interest in the colonies that was owned by a British company and place it either under the control of a statutory authority or transfer it to the Colonial Development Corporation or Overseas Food Corporation.[67] News circulated that Tate & Lyle was proposing to its shareholders that the firm sell off Caroni so this company would not be part of any nationalisation scheme.[68] Rumours of nationalisation reached St Kitts and the Colonial Office received a report that M. R. Bradshaw, president of St Kitts-Nevis Trades and Labour Union, had declared at a public meeting that he had received assurance from Arthur Creech Jones that the St Kitts Basse Terre sugar factory was to be nationalised.[69] Uncertainty about the future was also produced by the continuing strikes and acts of sabotage on many of the estates in Britain's Caribbean colonies, and relations between the sugar companies and their workforces were often very poor. Trinidad saw repeated unrest on its sugar estates in the post-war period, and a proliferation of unions made cooperation and consensus difficult to attain. We can also wonder about the discussions that seem likely to have occurred at firms such as Tate & Lyle about the fate of sugar operations once independence came for Britain's West Indian possessions. Although nationalisation did not become reality in the period before independence, and in the case of Trinidad it was not until 1970 that the government purchased the majority equity of Caroni, the threat of government control and the continuing labour discontent seems likely to have contributed to a sense of insecurity amongst sugar manufacturers that would not have encouraged them to expand their Caribbean activities in this period.

Economic conditions also militated against the success of new ventures in the British Caribbean using sugar as a raw material. The Barbados sugar cane wax factory was not the only enterprise using surplus sugar or molasses that failed to flourish. In 1950, a company producing alcohol, the Anhydrous Alcohol Company, was set up by the Sugar Manufacturers Association of Jamaica. The association had submitted a proposal to the Jamaican government to turn a surplus of molasses on the island into anhydrous alcohol that could be mixed with petrol and sold to consumers as a power alcohol. The project was pitched as a way to reduce unemployment and find an outlet for a by-product in excess. The Sugar Manufacturers Association argued that a reduction in imports of petrol would help the colony's balance of trade position and greater self-sufficiency in fuel was of benefit in the advent of war. There was also the prospect of using alcohol to establish other new industries such as plastics and paints. The Industrial Development Committee of Jamaica recommended the acceptance of the proposal for new legislation to make admixture compulsory up to 15 per cent, with an excise duty of 4d per gallon to be imposed and that the project receive Pioneer Industries status for four years.[70]

Before making his final decision, the Governor of Jamaica asked the Shell Company to solicit the views of experts on the use of power alcohols. Shell submitted a report on the performance of cars using petrol/alcohol blends in Cuba produced under a government scheme. This report suggested that the maximum alcohol content was around 15 or 16 per cent, at which point the anti-knock properties of the fuel improved, but beyond this, the acceleration of the car was affected.[71] DCL was also consulted and gave some technical advice, including the comment that the use of petrol/alcohol mixtures extended the life of an engine as the fuel gave smoother and cooler running.[72]

The Colonial Office was very positive about the development in Jamaica. Willis noted for his colleagues that the Jamaican sugar manufacturers aimed to dispose of 4 million gallons of molasses in their alcohol scheme.[73] The Economics Department commented that it fitted well with the more general aim of cultivating secondary industry in the colonies. At a meeting with officials in July 1950, Simonsen said that the question of the British Caribbean making power alcohol had come up in 1943 and whilst the conclusion at the time had been that the West Indies did not produce a sufficient volume of molasses to make a scheme viable, he supported the Jamaican proposal as it would release petrol for sale in dollar counties and the anti-knock properties of alcohol would reduce imports of lead tetraethyl that involved dollar expenditure. It would also bring experts into the colony that might help create other new chemical industry. Simonsen expressed his satisfaction that the creation of an alcohol factory in Jamaica was fulfilling the recommendations that he had made all along, saying in a letter to Eastwood, 'This is, of course, a policy which I have for a long time advocated for all sugar growing Colonies.'[74] The discussion of an alcohol industry in Jamaica resurrected other discussions about using the by-products of sugar cane. The potential of bagasse was brought up and a report produced that stated that there had been an investigation into the possibility of using bagasse in Trinidad to make wallboard and newspaper. For a period, the Trinidad-based subsidiary of Tate & Lyle, Caroni, had exported 3,000–4,000 tons of bagasse to the UK for use by the firm Celotex to make insulating board, but the firm was now reported to be changing over to straw. The economics of building a wallboard factory in Trinidad had been explored by Caroni and it had been concluded that the factory would be too small to be economic. The needs of the local market could be met in around two and half weeks of production and the high costs of freight for a heavy and bulky product prohibited its exportation. Additionally, Caroni had no success in interesting paper manufacturers in using bagasse for newsprint.[75]

The anhydrous alcohol project in Jamaica was approved by the legislature before suddenly coming under threat in 1953. The Jamaican government claimed to have received a 'disquieting report from one of the oil companies' that warned of serious technical problems, such as a tendency to vapour lock and engine corrosion. The opinion of DCL was again sought and the firm referred to the popularity of Discol during the interwar period with the remark, 'we could not make sufficient to meet public demand and they were quite prepared to pay 2d per gallon more for it than other nationally distributed fuel'.[76] By March, however, it was clear that the Jamaican government had changed its mind about the anhydrous alcohol scheme, stating that it could not bear the estimated loss of revenue of around £90,000 per year that would result from excise duty of only 4d per gallon each year for the first four years.

There was strong reaction to the news in the Jamaica newspaper, the *Daily Gleaner*, not least as the plant to make alcohol had already been constructed and production had begun only then for the government to rescind its agreement, seemingly after pressure from the oil companies. The newspaper linked the failure of the project to the closure of the food yeast factory, another enterprise that used surplus products from sugar manufacture. The newspaper claimed that the yeast factory provided an important source of protein for Jamaicans.[77] The Ministry of Food in Britain had created the Food Yeast Company Ltd and the Colonial Food Yeast Company in 1941 to manufacture edible food yeast. This strain of yeast had been originally developed by Thaysen when he worked at the DSIR's Chemical Research Laboratory at Teddington. The yeast was considered to have value on two counts – it grew on molasses and sugar, therefore providing a use for products in oversupply, and it had a high vitamin B and protein content and so could be a useful dietary supplement for Europe and the Far East during the war and, in the longer term, the colonies. The West Indian Sugar Company (a subsidiary of Tate & Lyle) were managing agents for the Colonial Food Yeast Company in the British West Indies and a factory was created next door to the WISCO estate at Frome using funds from the 1940 CDW Act. The operation of the factory was troubled, however, and the company barely made a profit and in some years made a loss. The price of sugar and molasses was higher than anticipated after the end of the war and the company also struggled to generate demand. For a period, the Malayan government bought a large amount of food yeast, but by the end of 1947 it had stopped. The Oxo Company placed one order but did not make another and the UN Relief Organization in Korea placed two orders but also then did not renew. Colonial governments were exhorted to purchase the yeast but few did. Trials were done in the British West Indian colonies in which straw-coloured flakes of dried yeast were incorporated into the menus provided in various types of institution – hospitals, dispensaries, the leper settlement of Antigua, schools in the Bahamas and Barbados, and also in gravies, stews and buns sold in British Guiana and Jamaica. Thaysen, the head of the CMRI, continued to advise on improving the process of food yeast manufacturing and recommended using it to fortify flour in Trinidad, reporting that sales of loaves from the island's bakeries had been excellent.[78] In general, however, the complaints were the same: the yeast was unpalatable and the price was too high. The CDC declined to take over the factory and recommended instead that WISCO take it on, but the UK government decided that selling a government-funded factory to a subsidiary of Tate & Lyle was not appropriate. By August 1952, it was decided to close the factory down.

The Jamaican *Gleaner* claimed that the failure of the venture was due to inadequate promotion, 'the truth is that a serious attempt has never been made to put the product on the world's markets. No commercial operator of such a venture would have relied on marketing arrangements as rudimentary as those attempted by the Food Yeast organization.'[79] The paper linked the termination of both the food yeast factory and the alcohol plant to a wider failure by government to instigate industrial development in Jamaica, 'In summary, in the midst of all the propaganda and drive and advertising for industrial development two important industrial schemes with which the Government is intimately connected are being scrapped. These two matters deserve immediate re-investigation.'[80]

The failures of sugar cane wax, levulinic acid, the food yeast factory and anhydrous alcohol show some of the obstacles facing new industry in the Caribbean using sugar. The costs of raw materials, building storage facilities and freight were all high and business struggled to find sufficiently large markets. New industry was highly dependent on government aid. The Pioneer Industries Legislation created in Jamaica, Trinidad and Barbados provided for income tax relief and the free import of machinery and helped support the sugar cane wax factory in Barbados for a period. This aid does not appear to have compensated for the problems that faced new industries, however, or was short lived, as in the case of Jamaica.

The CPC did not dwell on the wider economic and policy conditions that determined the success of sugar-based products or the fact that sugar manufacturers were not always committed to diversifying their interests. They focused their concerns on the work of the STL, particularly the notion that the researchers at the laboratory had not given sufficient consideration to the commercial viability of compounds and that its work had been too academic. Both the solution to the question of how to encourage new industry based on sugar, and then the criticism of that solution, were entirely focused on the operation of research.

To what extent was the CPC correct in seeing the fault as a failure of the STL to prioritise the commercial potential of products? The suggestion in some of the reports produced about the STL was that Wiggins lacked a certain industry-mindedness – that he was too academic. In fact, Wiggins was a product of Norman Haworth's laboratory at the University of Birmingham that had forged close ties with industry over decades. From 1926 Haworth was a consultant to ICI Dyestuffs in Blackley and remained so throughout his life, even beyond his formal retirement from Birmingham. One of Haworth's students, Maurice Stacey, said of Haworth, 'So many of his research students found their career at Blackley that at times some sections looked like an extramural arm of Birmingham University.' Apart from ICI, Haworth and Stacey were members of the consultancy team formed to advise Glaxo in the 1930s and the laboratory at Birmingham retained a close relationship with the firm.[81] Haworth was also a consultant from 1942 to the Birmingham Chemical Company of Lichfield and helped to found Nelson's Silk Co of Lancaster that manufactured cellulose acetate and employed many of Haworth's students, some of whom went on to form Nelsons Acetate Ltd in 1950.[82] As for Wiggins, on his resignation from the STL, he returned to Britain and took up the post of Research Director of the firm Aspro-Nicholas Ltd in Slough, a manufacturer of pharmaceutical, agricultural and household products.

The idea that scientists who advocated the idea of fundamental research into the chemistry of carbohydrates resided in an ivory tower far removed from the needs of industry does not appear to be an accurate description of the group at Birmingham that included Haworth, Stacey and Wiggins. Chemical firms in Britain often had close links with academic departments before the Second World War and the fact that many of Haworth's students went on to work at ICI and other smaller companies suggests the importance of informal networks that had taken years to develop. There may not have been formal apparatus in place in Britain to communicate needs to researchers and results to users but information passed between business and universities through informal, unrecorded or unpreserved communication on both sides. Similar contacts simply did not exist in the Caribbean. Wiggins appears to have tried to generate new networks by combining his duties as a researcher and laboratory director with time spent publicising the activities of the STL to attract new research fellows and funds, whilst also attending regional and international meetings and visiting companies. The research association formed was intended to cultivate a relationship between scientists and firms, but the reality was that the sugar companies had only ever agreed to participate in this arrangement reluctantly, and they lacked personnel with the right skills and knowledge to take up the discoveries made by Wiggins and his team. The reasons for the limited development of industry based on sugar in the region were far wider and more complex than any failure on the part of the researchers based at the STL.

Conclusion

Scientific research into colonial products had been promoted in the early 1940s as a way to deal with the problem of oversupply of commodities such as sugar on the world market. Frank Stockdale and Sydney Caine suggested that research was the key to establishing sugar as a raw material to make fuels and synthetic goods and diversifying the economies of the British West Indies. The search for new uses for sugar was done through in-depth analysis of the chemistry of this product and the elaboration of pathways to produce things like furfural, an intermediate used in the chemical industry to produce large numbers of different commercial products. In line with the wider approach to the organisation of scientific research at the time, scientists at Birmingham University and then the STL in Trinidad were given a great deal of freedom in determining their research programme.

The use of fundamental research as a synonym for freedom was a particular feature of the vision of scientific research that originated and was promoted by the research councils in Britain in the period after First World War. When representatives from these research councils were invited by the Colonial Office to create new arrangements for research in 1943, they promoted the idea of fundamental research into colonial problems and used this term to ensure that particular apparatus for research was created that provided for autonomy. Interestingly, scientists who came to the Colonial Office to join the CPC at the end of the 1940s did not show the same attachment to the activity of fundamental research and were more likely to assess research in terms of its usefulness than concern themselves with the need to ensure that researchers in the colonies were not in a subordinate position. There are a number of possible explanations for the criticisms that were made by Dobbs and Spensley about the work that had previously been done in the chemistry of colonial products. There is the possibility that the Colonial Office had become more circumspect in choosing its scientific advisors, taking the opportunity to engage scientists who were more likely to support the views of officials as they stood after 1947. It may be that scientists generally felt more confident about their professional standing by 1950 and so the concerns about the need to ensure a high status for research workers that had previously preoccupied individuals such as W. C. C. Topley and Edward Mellanby did not seem so important. It is also possible that the arrangements created for colonial research on the advice of representatives of the MRC, ARC and DSIR actually afforded research workers greater freedom in the colonies than scientists who worked at establishments at home. Spensley certainly believed that the STL did not operate as a laboratory run by a DSIR research association would do in Britain; that it was in fact more like a university department. Laboratories in Britain's colonies were more isolated than equivalent establishments in Britain, colonial governments often paid them very little attention and they were often very self-contained.

By the end of the 1940s, neither officials nor some of the scientists at the Colonial Office were convinced of the utility of fundamental research. As priorities at the Colonial Office and for Britain changed, scientific research into fields such as tropical products was reorganised to focus on more obvious practical issues. In particular, the idea of creating industry that used sugar as a raw material no longer seemed a realistic proposition by the mid 1950s. The era in which chemicals in Britain were made from coal or alcohol distilled from molasses ended with the Second World War. It had become apparent that the plans made by the Colonial Office to try to create a new future for the Caribbean in which sugar was reinvented as an industrial material coincided with the moment at which ICI and other companies were exploring a future based on oil. Sugar manufacturers were not interested in diversifying their work, and business ventures in the Caribbean suffered from a lack of government assistance in overcoming the problem of small local markets and expensive freight costs between islands and the South and North American mainland.

Notes

1 TNA, CO 852/1017/1.
2 Ibid.
3 TNA, CO 927/201/6.

4 *Ibid.*
5 *Ibid.*
6 S. Clarke, "The research council system and the politics of medical and agricultural research in the Colonial Empire, 1940–1952", *Medical History* 57 (2013), 338–358.
7 *Ibid.*
8 *Ibid.*
9 Lee and Petter, *The Colonial Office, War and Development Policy*, p. 171; Havinden and Meredith, *Colonialism and Development*, pp. 204–205.
10 Hyam, *The Labour Government and the End of Empire*, pp. xxiv–xxxi; Pearce, *Turning Point in Africa*.
11 TNA, CO 927/175/1.
12 TNA, CO 927/14/1.
13 Gummett, *Scientists in Whitehall*.
14 *The Colonial Empire (1947–1948)*, Cmd 7433; Hyam, *The Labour Government and the End of Empire*, pp. 42–45; TNA, CO 847/36/4.
15 TNA, CO 852/870/2.
16 TNA, CO 900/6.
17 Cantor, "Cortisone and the politics of empire", 463–493; Quirke, "Making *British* cortisone", 645–674.
18 CPRC *Fifth Annual Report, Colonial Research, 1947–1948*, Cmd 7493.
19 Committee for Colonial Agricultural, Animal Health and Forestry Research, *Third Annual Report, Colonial Research, 1947–1948*, Cmd 7493.
20 TNA, CO 852/1017/1; CO 927/201/6; CO 927/201/6.
21 At the same time, Dodds replaced Hankey as chair of the Biological Research Advisory Board that oversaw fermentation work at Porton Down.
22 TNA, CO 927/371.
23 *Colonial Research, 1947–1948*, Cmd 7493.
24 TNA, CO 899/5.
25 TNA, CO 899/6; British West Indies Sugar Association, *Annual Report*, 1957.
26 *Colonial Research, 1947–1948*, Cmd 7493.
27 *Colonial Research, 1949–1950*, Cmd 8063.
28 Maurice Stacey Archive (MSA), Birmingham University, B43/44.
29 TNA, FD23/1559.
30 *Colonial Products Research Council 5th Annual report, 1947–48.* Cmd 7493
31 TNA, FD 23/1559.
32 MSA, B43/44.
33 I. Zweiniger-Bargielowska, *Austerity in Britain: Rationing, Controls and Consumption 1939–1955* (Oxford: Oxford University Press, 2000), pp. 34–35.
34 TNA, FD 1/5960.
35 TNA, FD 23/1556.
36 J. Slinn and R. Davenport-Hines, *Glaxo: A History to 1962* (Cambridge: Cambridge University Press, 1992), p. 165.
37 TNA, CO 899/3.
38 C. Divall and S. F. Johnston, *Scaling Up: The Institution of Chemical Engineers and the Rise of a New Profession* (Dordrecht: Kluwer Academic Publishers, 2000), p. 155.
39 Reader, *Imperial Chemical Industries*, p. 401.
40 Divall and Johnston, *Scaling Up*, p. 156.
41 TNA, CO 927/371.
42 *Ibid.*
43 *Ibid.*
44 *Ibid.*
45 *Ibid.*
46 TNA, CO 899/6.
47 TNA, CO 913/3.
48 TNA, CO 911/5.
49 Pearce, *Turning Point in Africa*, pp. 61, 79, 85.

50 Pearce, *Turning Point in Africa*, Conclusion.
51 TNA, CO 927/371.
52 TNA, CO 900/5.
53 *Ibid.*
54 TNA, CO 899/4.
55 TNA, CO 899/4.
56 TNA, CO 927/406.
57 TNA, CO 899/5.
58 TNA, CO 899/5.
59 UWE, "British West Indies Sugar Research Scheme Annual Report 1953".
60 TNA, CO 899/5.
61 TNA, CO 899/4.
62 TNA, CO 899/5.
63 See, for example, *BWISA (Incorporated) Handbook for 1954* (Cambridge: W. Heffer and Sons).
64 TNA, CO 899/4.
65 D. Clampin and R. Noon, "The maverick Mr Cube: the resurgence of commercial marketing in postwar Britain", *Journal of Macromarketing* 31 (2011), 19–31.
66 TNA, CO 537/4509.
67 TNA, CO 5374509.
68 TNA, CO 537/4509.
69 TNA, CO 152/526/3.
70 TNA, CO 137/902/3.
71 TNA, CO 137/902/3.
72 TNA, CO 137/902/3.
73 TNA, CO 1031/1153.
74 TNA, CO 137/902/3.
75 TNA, CO 137/902/3.
76 TNA, CO 1031/1153.
77 TNA, CO 1031/1153.
78 TNA, CO 1029/47.
79 TNA, CO 1031/1153.
80 TNA, CO 1031/1153.
81 MSA, "Haworth Memorial Lecture. The Consequences of some projects initiated by Sir Norman Haworth", written for *Chemical Society Review*.
82 *Ibid.*

CONCLUSION

Science and industrial development: lessons from Britain's imperial past

For many places seeking to raise living standards after 1945, economic development came to mean industrialisation. By the 1950s, economists such as Raul Prebisch, Hans Singer and Paul Baran were advancing models of industrial development that promoted the necessity of restricting imports to allow new domestic industries to flourish. The 1950s also saw the rise of modernisation theory in which industrial revolution was central to the process of emerging as a modern state. This book has been concerned with development visions that were circulating before economists such as Prebisch and W. W. Rostow published their ideas. It has aimed to revise the usual story in which Britain resisted economic diversification in its Caribbean colonies and instead has shown that a number of visions of Caribbean industrialisation were proposed after 1942 that can be described as types of industrialisation-by-invitation, in a stronger or weaker form. These ideas, promoted by the British government, Arthur Lewis and the Caribbean Commission, differed so that no unified theory of development can be said to have informed plans for the Caribbean between 1940 and 1960. This account has explored the variety in proposals to encourage economic diversification that were expounded. The aim has been to demonstrate how different economic and political priorities worked to produce contrasting visions of industrial development, and explain in particular the emphasis on scientific research by the Colonial Office in London.

Amongst the various paths to industrialisation that were promoted to Caribbean legislatures after 1940, the least complete and coherent programme of economic diversification was that promoted by Britain. Officials in London favoured limited incentives from colonial governments for business and, in general, wished for minimal disturbance to market forces. Britain would not assume large financial risk in the process of establishing new business across the British Caribbean and it advised colonial governments to act similarly. In addition, officials did not believe that it was the job of planners appointed to an official body to determine the nature of new industry. The role of metropolitan government was to provide the money needed for the development of infrastructure as the necessary context for industrial development, and the provision of useful information. This included economic and business advice and also the funding of scientific research with the goal of opening up new opportunities for manufacturers to exploit tropical products.

The Colonial Office solution to the issue of how to facilitate industrial development in the British Caribbean was unique in giving a role to scientific research. Funds from the CDW Acts were used to sponsor laboratory investigations into cane sugar, with the aim of generating power alcohols, plastics and drugs that could be commercially produced. This was an ambitious plan in which the transformation of sugar into an industrial raw material would supposedly allow the British West Indies to escape the trap of being producers of low-value foodstuffs in oversupply. It was also a vision that confirmed the liberal values of the Economics Department of the Colonial Office. The Colonial Office committed funds for the scientific study of cane sugar on the basis that firms would take up the production of sugar-based compounds if the right sort of information was made available to them. This was a carefully delimited type of state intervention; government would assume responsibility for generating knowledge, but how that information was utilised was left up to business. This was an intervention intended to stimulate the creation of new industry in the Caribbean or encourage firms in the UK to take up new products that supposedly did not disturb the natural play of market forces. Decisions about the development of new products based on discoveries in state-funded laboratories resided with business. Firms would also provide the finance necessary to undertake this task. While officials were happy to support state-funded scientific research, they rejected the suggestion of a Caribbean development bank or regional development corporation with planning functions. For officials at the Colonial Office, the limited form of state intervention represented by scientific research was a relatively inexpensive and politically acceptable mode of state action to stimulate industrial development.

Sugar research was part of a wider commitment to funding scientific research in order to provide knowledge for colonial development after 1940. It is a commonplace claim in much scholarship on science and development to assert that the high profile given to scientists and technological solutions after 1945 resulted from an unquestioning faith in the superior nature of American and European science and technology, possibly fuelled by Allied success in the war and the emergence of technological wonder weapons such as penicillin, DDT and the atom bomb. In contrast to such claims, Britain's enthusiasm for investment in research as an activity essential for successful and effective colonial development cannot be reduced to some ill-defined enthusiasm for science in the 'atomic age'. The expansion of research in the Colonial Empire is better understood as one step in the more general rise of state-funded research over the first half of the twentieth century in Britain. Officials at the Colonial Office were not the first to formulate a relationship between scientific research and economic growth. The idea that the state should assume some responsibility for the production of knowledge relevant to issues of national importance – agriculture, health, industry and military matters – underpinned the creation and extension of government departments and laboratories that dealt with work in the civil and military spheres at home. It led to the formation of the research councils, beginning with the DSIR in 1916 and then continuing with the creation of the MRC in 1920 and ARC in 1933. It was the research council system that provided the model for colonial research after 1940. In the case of research into tropical products by the CPRC, the methods and rhetoric employed by the DSIR were of most importance. The CPRC, the DSIR and also the Development Commission and Empire Marketing Board offered something significant to politicians and officials concerned with economic policy, apart from the promise of useful facts. The creation of these agencies was facilitated by their political utility as these bodies provided a method of stimulating economic growth and development that was could be promoted as useful and effective, a demonstration of the willingness of government to take action, without being overly contentious or divisive. A commitment to funding scientific research was not associated with any party political or ideological position in the way that other more controversial types of government action to encourage economic and social change, such as the use of tariffs, subsidies or state-run companies and development banks, could be. In the vision of colonial development that emerged after 1940, scientific research could be invoked both for its compatibility with laissez-faire economics and also as the first stage of planning. The fact that scientific research could be associated with such a range of latent functions helps explain how agreement could be forged on the need for government to support this activity. Scientific research was a method of state intervention in economic matters around which political consensus could be reached. Research is not so much apolitical, but rather an activity with such a range of meanings and implications that it is flexible enough to be compatible with many political positions or requirements.

If the commitment of state funds to scientific research through the DSIR or CPRC had political expediency then this was because of the way that activity was characterised. Both bodies used the expression 'fundamental research' to describe the work they sponsored and this was often defined as enquiries into underlying phenomena or basic chemical reactions. The general nature of this work was important as it indicated that public money would advance a field to the benefit of a particular sector of industry and was not going to be spent on problems that were so practical they were close to the everyday problems of production, or so specific they concerned only individual manufacturers. It was important to avoid any accusation that state agencies allocated money that merely ended up lining the pockets of businessmen or favoured one firm over another. The meanings afforded to fundamental research between the First World War and the 1940s shaped a conception of state science that had a distinctly liberal character. The key value that was embodied in the state-funded research council system was freedom – for individual scientists to be free to choose their own research problems without direction, and also as the value underpinning the vision of how the results of research would benefit the nation or colony, namely that the responsibility of eminent research organisations was to ensure that scientific knowledge was freely available but not to engage in the process of uptake or application themselves. This was not the only vision of the relationship between scientific research and national needs that was available. We can compare the research council formulation of the place of scientific research in the progress of nation and empire with that of J. D. Bernal, for example, where 'planning' replaced 'freedom' as the cornerstone of the system. However, whilst the ideas of Bernal and other scientists on the Left have attracted much historical attention, the reality was that they were not significant in shaping the system of state-funded scientific research that emerged in Britain during the first half of the twentieth century. The research council model as promoted by individuals such as Edward Mellanby, W. W. C. Topley and A. V. Hill triumphed. The dominance of one particular vision of the way in which scientific research should be funded and organised, at a moment when government sought to find politically acceptable ways of increasing the capacity of the state to stimulate and manage change, was the particular accomplishment of those who worked in and defended the research council system, rather than something that was natural or inevitable. Important to the rise of the liberal ideology of research were claims about the supposed true nature of scientific research, often called fundamental research, by a group of scientists in Britain that sought to balance a desire to bring science and the state into a much closer relationship with the preservation of the professional freedoms and status of scientists.

Scientific research in Britain's tropical colonies, then, was not a thing apart but was shaped by the values and beliefs that had informed the metropolitan research system as it developed and which many scientists saw as having an important role in defending their interests, professional status and identity. The DSIR, MRC and ARC provided a model, and a defence, of particular arrangements for research in which the freedom for researchers to choose research problems for themselves was of paramount importance. The creation of the Research Department at the Colonial Office effectively created another research council to sit alongside the DSIR, MRC and ARC. In practice, this created a strange situation in which the Colonial Office mimicked the work of the research councils while still being a government department. This anomalous situation proved to be unsustainable in the long term, and by the 1950s, officials at the Colonial Office asserted that the priority in colonial research was for scientific research to meet colonial needs, rather than the preservation of the freedom for scientific researchers to determine their own research agenda.

Scientific research was funded for Britain's colonies on the basis that it might stimulate industrial development, amongst other things, and this study has considered the ensuing relationship between research and colonial development. The result has been to raise some questions about seemingly straightforward assertions about the role of state-funded scientific research as a motor for development and economic growth. One relates to the issue of the mechanisms or processes that allow knowledge to move from the laboratory and be translated into social and economic benefits. The fact that the Colonial Office and its scientific advisors did not much attend to the ways in which scientific discovery might be turned into commercial products can be explained by the use of the research system that operated in Britain as a model. The DSIR, MRC and ARC were concerned with projects of general fundamental research and it was said they left the development of any findings that arose through that work to other people. While research councils such as the MRC might have appeared to be only concerned with freely publicising the results of the work it sponsored, in fact researchers sponsored by the MRC were embedded in wider networks of pre-clinical and clinical researchers, government and business in Britain. These relationships were not necessarily formally defined but had developed over time in an ad hoc way through contacts that occurred at universities, clinics, London clubs, the numerous government committees that existed, and scientific societies. In the field of chemistry, research groups such as that headed by Norman Haworth at Birmingham University had a long history of acting as consultants to firms such as ICI and Glaxo. Students left the Chemistry Department at Birmingham and joined industry, or formed their own companies, and it seems highly likely that the most important way in which university–business contacts were deepened and maintained was through this employment of students who facilitated the exchange of knowledge and techniques between their old teachers and the firms they worked for. In Britain's colonies, scientists could become part of networks, but useful contacts that would facilitate the transfer of knowledge at the level of the individual colony did not necessarily have the time and opportunity to develop in the late colonial period.

Fundamental research into the chemistry of sugar was done on the basis that the results of scientific research would be of interest to businessmen. It became clear, however, that sugar manufacturers that operated in the Caribbean did not possess the necessary chemical and commercial skills to capitalise on the results presented to them. The sugar technologists who worked for firms such as Caroni in Trinidad focused their efforts on improving the efficiency and quality of technical processes in the sugar factory, rather than engaging in synthetic chemistry. This would seem to mirror the point that has been made in a recent study of the relationship between academic research and economic development in low- and middle-income countries. A report by UNESCO shows that university science has limited wider impact if there is a gap between this research and the technological and scientific know-how held in the private sector.[1] In other words, successful commercialisation requires a wider context in which a skilled workforce is employed by firms outside of the academy that can make use of university research findings, otherwise research may come to very little. That wider context was absent in the case of the research undertaken into sugar in Trinidad in the late colonial period. Although the sugar research scheme sponsored by the CPRC appears to have been a failure, it is worth remembering, however, that the findings of Wiggins and his team were subsequently taken up in countries such as Mauritius, where sugar was an important commodity, and have continued to inform research in the field of carbohydrate chemistry and the search for alternatives to petrol and new uses for sugar cane.[2]

Historical accounts that ask the question 'Why did development fail in the past?' have produced conclusions that run the gamut from the claim that scientists were not adequately consulted and so development failed because of insufficient research (see Havinden and Meredith on the Groundnut Scheme and the projects of the CDC),[3] through to the idea that scientific experts were given too central a role and so development failed because it dealt in overly simplified representations of nature and society. The contribution from this study is that the wider effects of scientific research are dependent upon other economic and institutional factors, such as human and financial capital, government assistance, markets and so on, to the extent that policies that focus on science are often missing the point. Tweaking science policy or increasing funding for science in African universities, for example, is not going to have the decisive effects that policy makers often claim. Scientific research can be important only if, and when, other conditions are in place.

It is highly unlikely that issues to do with the take-up of scientific results were the major factor that limited the industrial development of the British Caribbean. No amount of scientific research could overcome the problems of attracting investment for new industry to the region, the expense of shipping both between the British colonies and further afield, or the identification of adequate markets. While scientific research into finding new products and processes could play a part in facilitating industrial development, it could not be the motor. In order for scientific research to be important in economic development, certain preconditions needed to be met, such as the existence of industrial expertise, capital, services and incentives or support for new business. The policies of the Colonial Office were only marginally concerned with providing the wider context in which scientific research might have an impact. In contrast, the most fully worked-out visions of industrial development that were created for the Caribbean were much more transformative, seeking to create a wider set of conditions that would allow industry to flourish, including incentives for private investors through tax relief, trading estates, subsidies, cheap factories and credit. While the architects of these alternatives did not reject the need for scientific research useful to business, they did not make provision for it. The vision of industrial development promoted by W. Arthur Lewis, and representatives from Puerto Rico, was focused instead on the crucial role of foreign capital.

The most comprehensive plans for industrialisation, those of Lewis and Teodoro Moscoso of Puerto Rico, gave a bigger role to government in creating key industries, providing loans to businessmen and building factories. Additionally, officials on the American Section of the Caribbean Commission saw benefit in the idea of coordinated industrial development that involved some rationalisation of manufacturing across the Caribbean region, and trade agreements for the mutual support of the industrial activities that occurred in different territories. While the provision of advice was considered essential, and officials of the American Section wished to see an advisory body created to help plan and coordinate industrial development for the whole Caribbean, US representatives did not make plans to support scientific research in the way that the Colonial Office did.

Clearly, there were ideological differences between Arthur Shenfield and Sydney Caine at the Colonial Office on the one hand, and individuals such as Lewis and members of the American Section of the Caribbean Commission, on the other, in their expectations of the role of the state in economic affairs. These beliefs about the role of government in economic development were often contained in references to the speed and scale of change that was said to be appropriate or desirable. The fundamental differences between officials in London and those who favoured greater intervention for development was most apparent when the Colonial Office spoke of the need for gradual change, often naturally occurring, while others such as Lewis and officials from Puerto Rico claimed that what the Caribbean needed was a 'sudden jump' or rapid economic and social transformation. While Britain saw virtue in caution, US officials desired action and supported the more assertive approach voiced by representatives on the Caribbean Commission from Puerto Rico.

This investigation of the variety that existed in visions of industrialisation after 1940 leads us to a more fine-grained and nuanced understanding of the factors that shaped development thinking in the post-war period. It shows the limits of some of the broad descriptions of the nature of development that are used by scholars. The paths to industrial development that were promoted after 1940 cannot be explained merely by reference to the emergence of an orthodox development model. Ideas about development that were expressed during the 1940s and early 1950s were derived from various ideological, political and economic beliefs, practices or requirements. In this case, deeper consideration of political and economic pressures and ideals, beyond just a reference the emerging tensions of the Cold War, can help us understand the particularities of the visions of industrialisation that were promoted.

Differences in British and American priorities when it came to developing the Caribbean were related to the different function of development for those nations. US encouragement of industrial development in the Caribbean region was intended to bring prosperity to a region of strategic importance to America and also to further the wider aim of reducing barriers to trade around the world, including the opening-up of the colonies of the European powers to American firms. American officials wished to demonstrate the enlightened nature of US policies, and to do this it was imperative that they distinguish their actions from those of the European colonial powers. Such differentiation could be achieved by emphasising US commitment to the provision of political freedom for colonised peoples, amongst other things. While the Colonial Office worked to limit the powers of the Caribbean Commission, this body was a major instrument for the exercise of American power in the region. The US Section needed the Commission to be active and interventionist if it was to be proof that the US was serious in its commitment to improving the conditions that existed in poor nations in the post-war world, and if America was to shape that world in ways it considered to be politically and economically desirable.

The Colonial Office managed to resist most of the recommendations for a new Caribbean industrial policy that were made by the US Section of the Caribbean Commission during the 1940s and early 1950s. American ambitions for coordinated industrialisation, for trade agreements between territories, or for new regional agencies to fund and plan development, were all frustrated by opposition from Britain. For the Colonial Office, it was the CDW Acts, not the Caribbean Commission, that were intended to be an important demonstration of the altruism of British imperial rule. The CDW Acts would show colonial peoples, domestic audiences and critical foreign powers that British imperialism was not primarily concerned with exploitation but would bring social and economic benefits to the people that inhabited Britain's colonial possessions. The creation of laboratories for scientific research in the colonies had particular symbolic value after 1940 as concrete representations of Britain's commitment to modernising its possessions. National research laboratories could be presented as indicators of incipient modernity and the move towards an independent status. They brought to a colony the capacity both to be self-sufficient in knowledge and also to contribute to the international project of scientific advance.

The 1940 CDW Act was inspired by the experiences of the Great Depression that exposed both the fundamental weakness of colonial economies and the limited capacity of colonial governments to address those weaknesses. The recognition of the problems produced by the narrow base of primary products that underpinned colonial economies, and the growing problem of unemployment in some places, produced a turn towards industrial development. Historians have spoken of the way that the Colonial Office appointed expert metropolitan bodies, embraced planning and dispatched large numbers of new specialists to the Colonial Empire as part of an assertive, interventionist approach to colonial development during the early 1940s.

In practice, the Colonial Office approach to encouraging industrial development was decidedly more laissez-faire than that favoured by officials from Puerto Rico, some members of the US Department of State, Eric Williams and Arthur Lewis. The rather piecemeal and limited nature of British recommendations for the industrial development of the British Caribbean is likely to have contributed to their invisibility in the existing historical record. The principles that determined British attitudes towards industrial development for the Caribbean were very different from those discernible in British approaches to developing African agriculture. The work of Christophe Bonneuil, Dorothy Hodgson and others has focused on the sort of 'total' development scheme that colonial officers attempted to impose on African communities from the interwar period onwards. In projects such as the Gezira cotton-growing scheme in Sudan the aim was the complete restructuring of village life and work to turn subsistence farmers into efficient producers of cash crops. These projects could be very large and aimed to completely transform the lives of the African communities involved so that individuals became part of regimented and disciplined systems of production. In African colonies, then, the state could assume substantial responsibilities in the name of development, and employ methods that were highly interventionist, coercive and disruptive. Two related question emerge therefore – why were there such differences in approach to industrial development in the Caribbean between the Colonial Office and others, and why did the Colonial Office policies encompass such different conceptions of the role of the state in development when it came to industry and agriculture for Africa and the Caribbean?

The encouragement of large-scale agricultural production was a basic tenet of colonial policy, and agricultural development and the expansion of mining in Britain's colonies were effectively synonymous with economic development from the time of Joseph Chamberlain's tenure as Secretary of State for the Colonies in the 1890s. Colonial production of primary products was the basis of the notion of complementary economies in which Britain's tropical possessions provided the raw materials needed in Britain for consumption and manufacturing, and in doing so generated demand for the purchase of industrial goods from the metropolis. For officials, and the public more widely, agricultural development in Britain's possessions was an undeniable good; it was good for colonial producers and it was good for British workers and industrialists. This model of imperial economic relations did not readily accommodate the emergence of colonial industry. After 1940, economic diversification in the colonies was presented as an opportunity for British firms, but in reality industrial development was always in tension with the expectations of British manufacturers and labour. New colonial industry would most likely serve local markets, and in doing so it threatened to reduce demand for British products across the Colonial Empire. After 1945, the Colonial Office had to deal with the hostile reaction of British manufacturers to the news that special concessions were being made in some colonies for pioneer industries. A certain ambivalence towards the encouragement of colonial manufacturing on the part of officials, because of the possibility of complaints by British firms, is likely to have contributed to the conservative attitude of Colonial Office. More than this, however, industry was not considered essential for the colonies in the same way as agriculture, and therefore the issue of both the short- and long-term financial burden to government that might arise from the encouragement of colonial industry was a greater check on intervention. Colonial industry was desirable for the benefit it might bring in diversifying the base of colonial economies and relieving unemployment, but not at any cost. It would never displace the central place of agricultural production that was considered the natural and most suitable activity for Britain's tropical possessions. Sydney Caine was an advocate of industrialisation in the Caribbean but he did not believe that manufacturing would replace sugar production as the principal economic activity of Britain's West Indian territories. The Colonial Office attempted to rework and modernise the meanings of cane sugar through a search for new uses for this commodity as an industrial raw material in a vision of industrial development that aimed to prop up the ailing sugar industry rather than replace it. The point is that the sort of thoroughgoing, state-centred initiatives that were considered appropriate for the transformation of African farming practices were not needed in the case of industrial development, as industry was not considered central to the economic life of the colonies.

The Colonial Office ambition to foster some degree of industry in the colonies that was first expressed in 1943 was also frustrated in practice during the course of the 1940s as new demands were placed upon the Colonial Empire in line with changing domestic needs. Agricultural development schemes that were focused on increasing the production of cash crops for export were compatible with both the ambitions of the CDW Acts and also the requirement to improve the supply of foodstuffs and the dollar revenue available to Britain after 1947. While the currency controls and regulations on imports that were introduced to deal with Britain's economic crisis became a spur to increasing agricultural production, they were not compatible with the requirements of industrial development. Colonies such as Trinidad found their plans to attract new business frustrated by restrictions on imports of machinery from the US. Controls in place to conserve dollars and strengthen the sterling area made it hard for the colonies to make progress with industrial development, and officials in London did nothing to address the difficulties that faced colonial governments when it came to implementing their plans for industrialisation.

Altogether, the Colonial Office focus on industrialisation in the early 1940s was not quite the break with the past in practice that officials sometimes claimed. More generally, in fact, the new interventionist, active and constructive approach to development that historians have said was represented by the 1940 CDW Act did not amount to a turn towards detailed economic planning and the widespread assumption of new state functions. When Sydney Caine or Charles Carstairs spoke of the need for greater metropolitan initiative and clearer strategies for development by colonial governments, they were making an argument for the assumption of greater responsibility by the imperial and colonial state than had previously been assumed in colonial policy. Considering that in the past the primary task of colonial governments had been to balance the books – to spend little and collect taxes – this was not necessarily a grand vision of comprehensive economic planning. The challenge after 1945 was to reconcile an attachment to laissez-faire economics with a commitment to greater government-initiated change, in order to improve conditions in the British colonies and restore Britain's reputation as a colonial power. One way in which officials could resolve the potential tension between these imperatives is demonstrated by the role given to scientific research in Britain's programme for industrial development in the British Caribbean.

Notes

1 UNESCO, *Higher Education in Asia: Expanding Out, Expanding Up. The Rise of Graduate Education and University Research* (Montreal, Canada: UNESCO Institute for Statistics, 2014), ch. 3.
2 M. Patarau, *By-Products of the Cane Sugar Industry* (Amsterdam: Elsevier, 1989).
3 Havinden and Meredith, *Colonialism and Development*, Conclusion.

BIBLIOGRAPHY

Primary sources

Manuscript collections

The Barbados National Archive (BNA), Black Rock, Barbados.
Eric Williams Memorial Collection (EWMC), University of the West Indies, Trinidad.
Lewis Papers, Princeton University.
Library of the University of the West Indies, St Augustine Trinidad.
Maurice Stacey Archive (MSA), Birmingham University.
National Archives of Trinidad and Tobago (NATT), Port-of-Spain, Trinidad.
Rhodes House Library, University of Oxford.
The National Archives, London, United Kingdom.
National Records and Archives Administration (NARA), College Park, Maryland, USA.

Newspapers

The Canberra Times.
The Economist.
The Manchester Guardian.
Nature.
The Times.
The Trinidad Guardian.

Official publications

Colonial Research, 1942–1943, Cmd 6486.
Colonial Research, 1943–1944, Cmd 6535.
Colonial Research, 1944–1945, Cmd 6663.
Colonial Research, 1945–1946, Col. No. 208.
Colonial Research, 1946–1947, Cmd 7151.
Colonial Research, 1947–1948, Cmd 7493.
Colonial Research, 1948–1949, Cmd 7739.
Colonial Research, 1949–1950, Cmd 8063.
Colonial Research, 1950–1951, Cmd 8303.
Fuel Research Board, *Fuel for Motor Transport* (HMSO, 1921).
Fuel Research Board, *Fuel for Motor transport. An interim memorandum* (HMSO, 1920).
H. M. Petroleum Executive. *Report of the Inter-Departmental Committee on various matters concerning the production and utilisation of alcohol for power and traction purposes* (1919), Cmd 218.

BIBLIOGRAPHY

Report of the West Indian Conference held in Barbados 21-30 March, 1944, Colonial no. 187 (HMSO, 1944).
Report of the West India Royal Commission (1898), C.8655.
Report of the West India Royal Commission (1945) Cmd 6607.
Report of the West Indian Sugar Commission (1929), Cmd 3517.
Trinidad and Tobago Disturbances 1937: Report of the Commission (HMSO, 1938).

Other primary material

Beckles, W. A., *The Barbados Disturbances, 1937, Review – Reproduction of the Evidence and Report of the Commission, Bridgetown* (Barbados: Advocate Co, 1937).
BWISA (Inc), *Reports on Research Work 1943* (Barbados: Advocate Co, 1943).
BWISA (Incorporated) Handbook for 1954 (Cambridge: W. Heffer and Sons, 1954).
Chaston, P., "Gentlemanly Professionals Within the Civil Service: Scientists as Insiders During the Interwar Period" (DPhil, University of Kent at Canterbury: 1997).
Gomes, A., *Through a Maze of Colour* (Trinidad: Key Caribbean Publications, 1974).
Lewis, W. A., "Colonial Development", Lecture to Manchester Statistical Society (12 January 1949).
  Industrial Development in the Caribbean (Kent House, Port-of-Spain, Trinidad: Caribbean Commission Central Secretariat, 1951).
  "Economic development with unlimited supplies of labour", *The Manchester School* 22 (1954), 139–191.
Mellanby, E., *The State and Medical Research* (Edinburgh: Oliver and Boyd, 1939).
Scott, W., *The Industrial Utilisation of Sugar Cane By-Products* (Kent House, Port-of-Spain, Trinidad: Caribbean Commission Central Secretariat, 1950).
Shenfield, A. A., "Economic outlook, an upward trend", *New Commonwealth*, Special Caribbean Supplement 30 (1955), p. xi.
  "Economic advance in the West Indies", *New Commonwealth* 35 (1958), 357.
Taussig, C., "A program in the Caribbean", *Foreign Affairs* 24 (1946), 703.
Thaysen, A. C., 'Food yeast: its nutritive value and its production from Empire sources', *Journal of the Royal Society of the Arts* 93 (1950), 353–364.

Secondary sources

Anderson, D., "Depression, dust bowl, demography and drought: the colonial state and soil conservation in East Africa during the 1930s", *African Affairs*, 83 (1984), 321–343.
Anker, P., *Imperial Ecology: Environmental Order in the British Empire, 1895–1945* (Harvard: Harvard University Press, 2001).
Arndt, H. W., *Economic Development: The History of an Idea* (Chicago: The University of Chicago Press, 1987).
Ashton, S. R., and D. Killingray, *The West Indies*, British Documents on the End of Empire, Series B, 6 (London: HMSO, 1999).

BIBLIOGRAPHY

Ashton, S. R., and S. Stockwell, *Imperial Policy and Colonial Practice, 1925–1945*, British Documents on the End of Empire (London: HMSO, 1996).

Austoker, J., and L. Bryder (eds), *Historical Perspectives on the Role of the MRC: Essays in the History of the Medical Research Council of the United Kingdom and its Predecessor, the Medical Research Committee* (Oxford: Oxford University Press, 1989).

Balmer, B., *Britain and Biological Warfare: Expert Advice and Science Policy, 1930–65* (Basingstoke: Palgrave, 2001).

Beinart, W., and J. McGregor (eds), *Social History and African Environments* (Oxford: James Currey, 2003).

Bernal, R., "The Great Depression, colonial policy and industrialization in Jamaica", *Social and Economic Studies* 37(1/2) (1988), 33–64.

Bernal, V., "Colonial moral economy and the discipline of development: the Gezira scheme and 'Modern' Sudan", *Cultural Anthropology* 12 (1997), 447–479.

Bernton, H., W. Korarik and S. Sklar (eds), *The Forbidden Fuel: Power Alcohol in the Twentieth Century* (New York: B. Griffin, 1982).

van Beusekom, M. M., *Negotiating Development: African Farmers and Colonial Experts at the Office du Niger, 1920–1960* (Westport: Heinemann, 2002).

Bolland, N., *On the March: Labour Rebellions in the British Caribbean, 1934–39* (Kingston: Ian Randle, 1995).

The Politics of Labour in the British Caribbean: The Social Origins of Authoritarianism and Democracy in the Labour Movement (Kingston: Ian Randle, 2001).

Bonneuil, C., "Development as experiment: science and state-building in late colonial and post-colonial Africa, 1930–1970", *Osiris* 15 (2000), 1501–1520.

Boodhoo, K. I., *Eric Williams: The Man and the Leader* (Maryland: University Press of America, 1986).

Brereton, B., *A History of Modern Trinidad, 1783–1962* (London: Heinemann, 1983).

Brown, J. M., and Wm. Roger Louis (eds), *The Oxford History of the British Empire*, Vol. VI, The Twentieth Century (Oxford: Oxford University Press, 2001).

Butler, L. J., *Industrialisation and the British Colonial State: West Africa, 1939–1951* (London: Frank Cass, 1997).

Cantor, D., "Cortisone and the politics of empire: imperialism and British medicine, 1918–1955", *Bulletin of the History of Medicine* 67 (1993), 463–493.

Carrington, E., "The post-war political economy of Trinidad and Tobago", *New World Quarterly* 4 (1967), 130–145.

Cell, J., *Hailey, A Study in British Imperialism, 1872–1969* (Cambridge: Cambridge University Press, 1992).

Chalmin, P., *Tate and Lyle: The Making of a Sugar Giant, 1859–1989* (London: Routledge, 1990).

Clampin, D., and R. Noon, "The maverick Mr Cube: the resurgence of commercial marketing in postwar Britain", *Journal of Macromarketing*, 31 (2011), 19–31.

Clarke, S., "A technocratic imperial state? The Colonial Office and scientific research, 1940–1960", *Twentieth Century British History* 18(4) (2007), 453–480.

Clarke, S., "Pure science with a practical aim: the meanings of 'fundamental research' in Britain, c1916–1950," *Isis* 101 (2010), 285–311.

"The research council system and the politics of medical and agricultural research in the Colonial Empire, 1940–1952", *Medical History* 57 (2013), 338–358.

Constantine, S., *The Making of British Colonial Development Policy* (London: Maurice Temple Smith, 1984).

Buy and Build: The Advertising Posters of the Empire Marketing Board (London: HMSO, 1986).

Cooper, F., and R. Packard (eds), *International Development and the Social Sciences: Essays on the History and Politics of Knowledge* (Berkeley and Los Angeles: University of California Press, 1997).

Cramer, L. W., "Foreword", *Industrial Development in the Caribbean* (Port-of-Spain, Trinidad: Caribbean Commission Central Secretariat, Kent House, 1951).

Curtin, P. D., "The British sugar duties and West Indian prosperity", *The Journal of Economic History* 14 (1954), 157–164.

Davies, R., "The rise of protection in England, 1689–1786", *The Economic History Review* 19 (1966), 306–317.

Deerr, N., *The History of Sugar*, Vol. II (London: Chapman & Hall, 1950).

DeJager, T., "Pure science and practical interests: the origins of the Agricultural Research Council, 1930–1937", *Minerva* 31 (1993), 129–140.

Divall, C., and S. F. Johnston, *Scaling Up: The Institution of Chemical Engineers and the Rise of a New Profession* (Dordrecht: Kluwer Academic Publishers, 2000).

Drayton, R., *Nature's Government: Science, Imperial Britain and the "Improvement" of the World* (London: Yale University Press, 2000).

Edgerton, D., *Warfare State: Britain, 1920–1970* (Cambridge: Cambridge University Press, 2005).

Britain's War Machine: Weapons, Resources and Experts in the Second World War (Oxford: Oxford University Press, 2011).

Ertsen, M. W., *Improvising Planned Development on the Gezira Plain, 1900–1980* (New York: Palgrave Macmillan, 2016).

Farnie, D. A., "The commercial empire of the Atlantic, 1607–1783", *The Economic History Review* 15 (1962), 205–218.

Farrell, T., "Arthur Lewis and the case for Caribbean industrialisation", *Social and Economic Studies* 29(4), 52–75.

Ferguson, J., *The Anti-Politics Machine: "Development", "Depoliticization" and Bureaucratic Power in Lesotho* (Cambridge: Cambridge University Press, 1990).

Figueroa, M., "The academic economist as public teacher: lessons from W. Arthur Lewis and the Caribbean Policy Discourse", *Social and Economic Studies* 58 (2009), 1–28.

BIBLIOGRAPHY

Fraser, C., *Ambivalent Anti-Colonialism: The United States and the Genesis of West Indian Independence, 1940–1964* (Westport: Greenwood Press, 1994).

French, J., "Colonial policy towards women after the 1938 uprising; the case of Jamaica", *Caribbean Quarterly* 34 (1988), 38–61.

Galloway, J. H., "Botany in the service of empire: the Barbados cane-breeding program and the revival of the Caribbean sugar industry, 1880s-1930s", *Annals of the Association of American Geographers* 86 (1996), 682–706.

Goldsworthy, D., *Colonial Issues in British Politics, 1945–1961* (Oxford: Oxford University Press, 1971).

Greene, J. P., "Society and economy in the British Caribbean during the seventeenth and eighteenth centuries", *The American Historical Review* 79 (1974), 1499–1517.

Griffith, W. H., "Lewis and Caribbean industrialization: policy, theory and the new technology", *The Journal of Developing Areas* 25 (1991), 207–230.

Gummett, P., *Scientists in Whitehall* (Manchester: Manchester University Press, 1980).

Haber, L. F., *The Chemical Industry, 1900–1930: International Growth and Technological Change* (Oxford: Clarendon Press, 1971).

Hailey, M., *An African Survey* (Oxford: Oxford University Press, 1938).

Harris, R., "Making leeway in the Leewards, 1929–51: the negotiation of colonial development", *The Journal of Imperial and Commonwealth History* 33 (2005), 393–418.

"From miser to spendthrift: public housing and the vulnerability of colonialism in Barbados, 1935–1965", *Journal of Urban History* 33 (2007), 443–466.

Harrison, M., *Jamaica, the Caribbean and the World Sugar Industry* (New York: New York University Press, 2001).

Hastings, J. J., "Development of the fermentation industries in Great Britain", *Advances in Applied Microbiology* 14 (1971), 1–45.

Havinden, M., and D. Meredith, *Colonialism and Development: Britain and Its Tropical Colonies, 1850–1960* (London and New York: Routledge, 1993).

Hecht, G. (ed.), *Entangled Geographies: Empire and Technopolitics in the Global Cold War* (Cambridge, Mass.: MIT Press, 2011).

Hoag, H. J., and M. B. Ohman, "Turning water into power: debates over the development of Tanzania's Rufiji River Basin, 1945–1985", *Technology and Culture* 49 (2008), 624–651.

Hodge, J., "Science, development and empire: the Colonial Advisory Council on Agriculture and Animal Health", *Journal of Imperial and Commonwealth History* 30 (2002), 1–26.

Triumph of the Expert: Agrarian Doctrines of Development and the Legacies of British Colonialism (Ohio: Ohio University Press, 2007).

Hodge, J. M., G. Hodl, and M. Kopf, *Developing Africa: Concepts and Practices in Twentieth-Century Colonialism* (Manchester: Manchester University Press, 2014).

BIBLIOGRAPHY

Hoppe, K. A., *Lords of the Fly: Sleeping Sickness Control in British East Africa, 1900–1960* (Westport: Praeger, 2003).

Horrocks, S., "Industrial chemistry and its changing patrons at the University of Liverpool, 1926–1951", *Technology and Culture* 48 (2007), 43–66.

Hull, A., "War of words: the public science of the British scientific community and the origins of the Department of Scientific and Industrial Research, 1914–16", *British Journal for the History of Science* 32 (1999), 461–481.

Hyam, R., *The Labour Government and the End of Empire, 1945–1951* (London: HMSO, 1992).

Ingham, B., "Shaping opinion on development policy: economists at the Colonial Office during World War II", *History of Political Economy* 24 (1992), 689–710.

Jasanoff, S. (ed.), *States of Development: The Co-Production of Science and Social Order* (London: Routledge, 2004).

Jeffries, C., *A Review of Colonial Research, 1940–1960* (London: HMSO, 1964).

Johnson, H., "The West Indies and the conversion of the British official classes to the development idea", *The Journal of Commonwealth and Comparative Politics* 15 (1977), 55–83.

Jones, M., "A 'Textbook Pattern'? Malaria control and eradication in Jamaica, 1910–1965", *Medical History* 57(3) (2013), 397–419.

Kelleher Storey, W., *Science and Power in Colonial Mauritius* (Rochester: Rochester University Press, 1997).

Kiely, R., *The Politics of Labour and Development in Trinidad* (Kingston: The University of the West Indies Press, 1996).

Kiely, R., *Industrialisation and Development: A Comparative Analysis* (London: UCL Press, 1998).

King, A., *Science and Policy: The International Stimulus* (Oxford: Oxford University Press, 1984).

Kumar, P., "Plantation science: improving natural indigo in Colonial India, 1860–1913", *British Journal for the History of Science* 40 (2007), 532–553.

Lauriat, B., "'The examination of everything': Royal Commissions in British legal history", *Statute Law Review* 31 (2010), 24–46.

Lee, J. M., and M. Petter, *The Colonial Office, War and Development Policy: Organisation and the Planning of a Metropolitan Initiative, 1939–1945* (London: Maurice Temple Smith, 1982).

Mackenzie, J. M. (ed.), *Imperialism and the Natural World* (Manchester: Manchester University Press, 1990).

Maldonado, A. W., *Teodoro Moscoso and Puerto Rico's Operation Bootstrap* (Gainesville: University Press of Florida, 1997).

Masefield, G. B., *A History of the Colonial Agricultural Service* (Oxford: Clarendon Press, 1972).

McCracken, J., "Experts and expertise in Colonial Malawi", *African Affairs* 81 (1982), 101–116.

Meighoo, K., *Politics in a Half-Made Society: Trinidad and Tobago 1925–2001* (Kingston: Ian Randle, 2003).

Meneight, W. A., *A History of the United Molasses Company Ltd* (Liverpool: Seel House Press, 1977).

BIBLIOGRAPHY

Morgan, D. J., *The Official History of Colonial Development* (London: Macmillan, 1980).

Morgan, K., *Slavery, Atlantic Trade and the British Economy, 1660–1800* (Cambridge: Cambridge University Press, 2000).

Morris, P., *The American Synthetic Rubber Research Programme* (Philadelphia: University of Pennsylvania Press, 1989).

Neill, D., *Networks in Tropical Medicine: Internationalism, Colonialism and the Rise of a Medical Specialty, 1890–1930* (Stanford: Stanford University Press, 2012).

Newbury, C., "Trade and technology in West Africa: the case of the Niger Company, 1900–1920", *The Journal of African History* 19 (1978), 551–575.

Oxaal, I., *Black Intellectuals Come to Power: The Rise of Creole Nationalism in Trinidad and Tobago* (Cambridge, Mass.: Schenkman Publishing, 1968).

Pantin, D. (ed.), *The Caribbean Economy: A Reader* (Jamaica: Ian Randle, 2005).

Pantojas-Garcia, E., *Development Strategies as Ideology: Puerto Rico's Export-Led Industrialisation Experience* (Boulder: Lynne Rienner, 1990).

Parker, J., "'Capital of the Caribbean': the African American–West Indian 'Harlem Nexus' and the Transnational Drive for Black Freedom, 1940–1948", *The Journal of African American History* 89 (2004), 98–117.

Brother's Keeper: The United States, Race and Empire in the British Caribbean, 1937–1962 (Oxford: Oxford University Press, 2008).

Patarau, M., *By-Products of the Cane Sugar Industry* (Amsterdam: Elsevier, 1989).

Payne, A., and P. Sutton, *Charting Caribbean Development* (London: Macmillan Caribbean, 2001).

Pearce, R. D., *The Turning Point in Africa: British Colonial Policy, 1938–1948* (London: Frank Cass, 1982).

Petter, M., "Sir Sydney Caine and the Colonial Office in the Second World War: a career in the making", *Canadian Journal of History* 16 (1981), 68–85.

Poole, J. B., and K. Andrews, *The Government of Science in Britain* (London: Weidenfeld & Nicolson, 1972).

Quirke, V., "Making *British* cortisone: Glaxo and the development of corticosteroids in Britain in the 1950s-1960s", *Studies in the History and Philosophy of Biological and Biomedical Sciences* 36 (2005), 645–674.

Reader, W., *Imperial Chemical Industries: A History* (Oxford: Oxford University Press, 1970), pp. 322–323.

Reinharz, J., "Science in the service of politics: the case of Chaim Weizmann during the First World War", *The English Historical Review* 100 (1985), 572–603.

Reynolds, L. A., and E. M. Tansey (eds), "British contributions to medical research and education in Africa after the Second World War", *Wellcome Witnesses to the Twentieth Century* 10 (2001).

Richardson, B., "Depression riots and the calling of the 1897 West India Royal Commission", *New West Indian Guide* 66 (1992), 169–191.

Roberts, G., and A. Simmonds, "British chemists abroad, 1887–1971: the dynamics of chemists' careers", *Annals of Science* 66 (2009), 103–128.

Roger Louis, Wm., *Imperialism at Bay, 1941–1945: The United States and the Decolonisation of the British Empire* (Oxford: Oxford University Press, 1977).

Rollings, N., "Cracks in the post-war Keynesian settlement? The role of organized business in Britain in the rise of neoliberalism before Margaret Thatcher", *Twentieth Century British History* 24 (2013), 637–659.

Rose, S., and H. Rose, *Science and Society* (London: Penguin, 1969).

Roskill, S., *Hankey: Man of Secrets* (London: HarperCollins, 1974).

Rist, G., *The History of Development: From Western Origins to Global Faith* (New York: Zed Books, 1999).

Ryan, S., *Race and Nationalism in Trinidad and Tobago: A Study of Decolonization in a Multi-Racial Society* (Toronto: University of Toronto Press, 1972).

Saul, S. B., "The economic significance of 'constructive imperialism'", *The Journal of Economic History* 17 (1957), 173–192.

Scott, J., *Seeing Like a State: How Certain Schemes to Improve the Human Condition Have Failed* (Yale: Yale University Press, 1998).

Sheridan, R. B., "The Molasses Act and the market strategy of the British sugar planters", *The Journal of Economic History* 17 (1957), 69–72.

Showers, K., "Soil erosion in the Kingdom of Lesotho: origins and colonial response, 1830s–1950s", *Journal of Southern African Studies* 15 (1989), 263–286.

Slinn, J., and R. Davenport-Hines, *Glaxo: A History to 1962* (Cambridge: Cambridge University Press, 1992).

Smith, S. D., "Merchants and planters revisited", *The Economic History Review* 55 (2002), 434–465.

Staples, A., *The Birth of Development: How the World Bank, Food and Agriculture Organization, and World Health Organization Changed the World, 1945–1965* (Ohio: Kent State University Press, 2006).

datt Tewarie, B., and R. Hosein, *Trade Investment and Development in the Contemporary Caribbean* (Kingston: Ian Randle, 2007).

Thomas, M., *Violence and Colonial Order: Police, Workers and Protest in the European Colonial Empires, 1918–1940* (Cambridge: Cambridge University Press, 2012).

Tilley, H., *Africa as a Living Laboratory: Empire, Development and the Problem of Scientific Knowledge, 1870–1950* (Chicago: The University of Chicago Press, 2011).

Travis, A. S., H. G. Schroter, E. Homburg and P. Morris (eds), *Determinants in the Evolution of the European Chemical Industry, 1900–193* (Dordrecht: Springer, 1998).

UNESCO, *Higher Education in Asia: Expanding Out, Expanding Up. The Rise of Graduate Education and University Research* (Montreal, Canada: UNESCO Institute for Statistics, 2014).

Varcoe, I., "Co-operative research associations in British industry, 1918–1934", *Minerva* 19 (1981), 433–463.

Vernon, K., "Microbes at work: micro-organisms, the DSIR and industry in Britain, 1900–1936", *Annals of Science* 51 (1994), 593–613.

BIBLIOGRAPHY

Vig, N. J., *Science and Technology in British Politics* (Oxford: Pergamon Press, 1968).

Ward, J. R., "The profitability of sugar planting in the British West Indies, 1650–1834", *Economic History Review* 31 (1978), 197–213.

Weir, R., *The History of the Distillers Company: Diversification and Growth in Whiskey and Chemicals* (Oxford: Clarendon Press, 1995).

Whitham, C., *Bitter Rehearsal: British and American Planning for a Post-War West Indies* (Westport: Praeger, 2002).

Wilkie, T., *British Science and Politics Since 1945* (Oxford: Basil Blackwell, 1991).

Will, H. A., "Colonial policy and economic development in the British West Indies, 1895–1903" *The Economic History Review* 23 (1970), 129–147.

Worthington, E. B., *Science in Africa: A Review of Scientific Research Relating to Tropical and Southern Africa* (Oxford: Oxford University Press, 1938).

Worboys, M., "Science and British Colonial Imperialism, 1895–1940" (DPhil, University of Sussex, 1979).

"The Imperial Institute: the state and the development of the natural resources of the colonial empire, 1887–1923", in J. M. Mackenzie (ed.), *Imperialism and the Natural World* (Manchester: Manchester University Press, 1990), pp. 164–186.

Zweiniger-Bargielowska, I., *Austerity in Britain: Rationing, Controls and Consumption 1939–1955* (Oxford: Oxford University Press, 2000).

INDEX

acetone 38–40, 164
Africa 5, 7, 8, 11, 50, 159, 190
   development programmes 2, 6
   research institutions 51, 60, 61
African Research Survey 5
Agricultural Research Council (ARC) 7, 9, 53–56, 64, 156, 183
agriculture 4–5, 24, 51–54, 60–61, 109, 190
alcohol 36–41, 65, 106, 107, 108, 146, 164, 172–173
Amani, Tanganyika 51, 109, 123
American bases 80, 132–133
An African Survey 53
Anglo-American Caribbean Commission
   see also Caribbean Commission 79, 80
Anglo-Iranian Oil Company 164
antibiotics 65, 111, 119, 121, 125, 171
Argus Chemical Corporation 114, 163, 168
Aspro-Nicholas Ltd 176
aviation fuel 32, 37, 42–43, 133, 164

Bacteriological Research Advisory Board (BRAB) 65
bagasse 114, 173–174
bananas 13, 25, 119, 121, 137
Barbados 12–13, 15, 23, 27, 92
   1937 riots 28–30
   sugar breeding 24–25
   sugar cane wax factory 115, 118, 146, 168, 170, 172, 175
Barclays Bank 93
Barlow Report 59, 64
Best, Lloyd 149
Birmingham University 41–42, 64–67, 111, 116, 162–163, 176, 186
blood plasma 163–164

Booker Bros 26, 115, 117
British West Indies Sugar Association (BWISA) 110–114, 118, 126, 168–170
Burgess, C. J. 95–96, 140–143
businessmen 62, 76, 88, 93, 105, 139, 143–147
Butler, Uriah 28–29, 132–134

Caine, Sydney 25, 35–36, 41, 62, 76–79, 85–89, 159, 171, 188, 191–192
Caribbean
   exports 12, 13–14, 25, 35, 133
   independence 14, 21, 90, 95, 148, 158, 160, 169, 172
   political reform 3, 14, 82, 90, 131
Caribbean Commission 15–16, 79–84, 89–97, 118, 131, 140–141, 147–149, 155, 181, 187–189
Caribbean Development Corporation 92, 13
Caribbean Economic Review 81, 97, 140
Caroni 172–174, 186
Carrington, Edwin 134, 148–149
Carstairs, Charles 55, 123, 157–158
Chamberlain, Joseph 4, 24, 190
chemical industry 16, 36–44, 65–66, 107, 110, 171, 173
chemurgy 37
Clauson, Gerald 37, 41, 65, 123
Clifford, Bede 88
coal 37, 41, 65, 67, 107, 178
cocoa 13–14, 25, 28, 69, 109, 119, 125, 135, 168
Colonial Agricultural Service 53, 60, 109
colonial development 33–34, 49–57, 60, 119, 156, 159, 182, 185, 189

[202]

INDEX

Colonial Development Act (1929) 5, 33, 50–51
Colonial Development Corporation (CDC) 83, 92, 93, 172, 175, 186
Colonial Development and Welfare Act (1940, 1945 CDW Act) 3, 7, 9–11, 26, 32–34, 41, 44, 49–52, 55–56, 62, 70–71, 76–78, 84, 110–111, 117, 122–123, 131, 154, 160, 166–167, 174, 189–191
Colonial Development and Welfare Organisation (CDW Org) 34, 84, 94–96, 121, 132, 146
Colonial Economic Advisory Committee (CEAC) 78–79, 84–87, 97, 113
Colonial Food Yeast Company 120, 125, 174
Colonial Medical Research Committee 57, 158
Colonial Microbiological Research Institute (CMRI) 104–105, 119–126, 154, 167–171, 175
Colonial Office
  advisors 53, 56–60, 64, 84–87
  Economics Department 32, 34–37, 40, 44, 55, 87, 101, 113, 173, 182
  Research Department 55, 123, 157, 158, 185
Colonial Products Laboratory 154, 165, 168
Colonial Products Research Council (CPRC) 41, 43, 62–70, 104–126, 146, 154–158, 161–167, 171, 183, 184, 186
colonial services 8, 53–57, 59–61, 63
Colonial Research Committee (CRC) 55–62, 78, 159
comirin 125, 170
commonwealth 90, 95, 122, 124, 158–159
communists 27, 31
cordite 38
cortisone 69, 161

Cramer, Lawrence 94, 97, 141
Creech Jones, Arthur 132, 158, 159, 172
customs union 91, 98, 142

Daily Gleaner 174, 175
decolonisation 76, 166–162, 167
Department of Scientific and Industrial Research (DSIR) 9, 37, 53–56, 59, 64–69, 110–113, 120, 126, 156, 165–166, 169, 174, 178, 183–185
development
  history 2–6
  planning in Trinidad 130–132
  see also colonial development
dextran 162–164
Discol 40, 174
Distillers Company Limited (DCL) 36, 40–41, 106–107, 118, 164, 173–174
diversification 83, 101, 105–106, 126, 132, 147, 181, 190
Dodds, Charles 162, 165, 170
Durbin, Evan 78–79, 85–89

economists 78–79, 85–89, 92–94, 96–99, 130–132, 135–143, 148–149
emancipation 23
Empire Marketing Board (EMB) 63
Ethiopia 28, 29
Ethylene 40, 107, 108, 162, 171
eugenol 67, 162

Fauvel, Luc 92–94
Federation 14, 90–91, 98, 142, 150
fermentation 37–39, 67, 108, 119–122
First World War 5, 25, 27, 29, 36–38, 43–44, 59, 177
food yeast 120, 121, 125, 161, 168, 174–175
forest products 69, 162
France 37, 81, 90, 92
Fuel Research Board (FRB) 37–39
fuels 15, 40–44

INDEX

fundamental research 9–10, 16–17, 50, 58–64, 67–70, 104–105, 112–115, 125–126, 155–158, 167, 176–178, 184–186
furfural 66, 114, 168, 177

Galletti, Robert 93
Galley, R. A. E. 162, 170
Germany 24, 31, 34, 51
Glaxo 161, 164, 176
Gomes, Albert 131–149
Great Depression 12, 22, 25–27, 33, 36, 37, 51, 77, 90, 99, 109
Griffith, James 136

Hailey, Malcolm 52–56
Hankey, Maurice 43, 56, 64–65, 112, 119–120, 125, 158, 162–163
Haworth, Norman 41–43, 64–66, 111, 162–163, 171, 176, 186
Heesterman, J. E. 94
Heilbron I. 64, 65, 67, 162
Hibbert, J. G. 157–165
Hilditch, T. P. 68
Hobson, Patrick 147–148
housing 30–31, 81, 133, 148

ICI 36, 40, 65–66, 69, 106–107, 144, 164
Imperial College of Tropical Agriculture (ICTA) 109–118, 121, 167–170
Imperial Department of Agriculture, Barbados 24
Imperial Institute 62–63, 65, 68, 157, 166
imperial preference 25, 63, 91–92
Inconvenience Allowance 36, 38, 41, 107–108, 164
industrial development bank 81, 83, 89, 92–93, 99, 126, 141–143, 182–183
Industrial Development Conference, Puerto Rico, (February 1952) 97, 140–143
industrial development corporation 98–99, 139, 142, 144, 147

industrial estates 86, 98, 136, 142–144, 148, 187
industrial survey
   British business leaders 143–146
   Caribbean Commission 92–94
   W. Arthur Lewis 96–99, 140–143
   Trinidad 130–136
industrialisation-by-invitation 97, 148, 181
industry
   cement 88–89, 92, 136
   Colonial Office policy 76–79, 84–89
   debate at Caribbean Commission 89–99
   Trinidad policy 135–140, 147–149
   Puerto Rico 81–83, 140–141
International Sugar Agreement 26, 34, 111

Jamaica
   1938 riots 31
   anhydrous alcohol factory 146, 172–175
   Industrial Development Corporation 139, 144, 147
   universal franchise 14, 80

laboratories 3, 9, 60, 104–105, 111–125, 167–171, 178, 182–183
laissez-faire economics 2, 6, 79, 87–89, 98, 101, 113, 159, 183, 192
levulinic acid 66, 111, 118, 162–163, 168, 170, 175
Lewis, W. A.
   at the Colonial Office 78–79, 85–87
   Caribbean Commission 96–97, 98–99, 140–143
   'The industrialisation of the British West Indies' 98–99, 140–143

MacDonald, Malcolm 21, 33–34, 52–53
Medical Research Council (MRC) 7, 9, 51–58, 156, 158, 161–166, 171, 183, 185

[204]

INDEX

Mellanby, Edward 54–56, 178, 184
microbiology 38–39, 104, 119–125
Ministry of Food 34, 108, 172, 174
Ministry of Supply 163, 164
modernisation theory 2, 76, 181
molasses 36, 39–41, 43, 67, 80, 106–108, 114, 116, 120, 146, 164–165, 168, 172–174
Morris, Daniel 25
Moscoso, Teodoro 81–84, 91–92, 97, 187
Moyne Commission 15, 21–22, 32–36, 43–44, 80, 129

nationalisation 149, 172
Navigation Acts 23
Netherlands 81, 92, 124
New Commonwealth 138
New Deal 79–82, 99
New World Group 149
Norman Commission 24
Northern Regional Research Laboratory, Peoria 38, 109, 114, 120–122

oil 12, 21, 27, 28, 30, 36–37, 40–43, 65, 107–109, 116, 132–136, 149, 164, 171, 174, 178
Olivier, Sydney 25–26, 35
Operation Bootstrap 82, 96, 97, 140–150

Panama Canal 12, 27
Payne, Clement 28–30
Peoples' National Movement (PNM) 147–149
penicillin 38, 114, 119, 121
Pioneer Industries 98, 115, 118, 129–131, 136–149, 168, 173–174, 191
planning 3–4, 16, 50, 55, 62, 78–79, 87, 90, 93, 105, 132, 135, 158, 183–184, 189, 192
plastics 16, 36, 22, 36, 40, 49, 64, 65–66, 106, 111, 114, 121, 163, 173, 182
Point Four 76, 95

Porton Down 65, 120
power alcohol 36–40, 168, 172–173
PRIDCO 81–83, 97, 137, 141
Privy Council 54, 64
Puerto Rico 80–83, 91–92, 96–99, 117, 130, 137, 140–150, 168, 187–190

quotas 26, 110–111, 133, 149, 137, 144, 164

Rance, Hubert 139
raw materials 22, 32, 36, 39, 40, 41, 42, 44, 64, 65, 107, 110, 112, 118, 146, 148, 163–164, 171–175, 177, 182, 191
research council 7, 9–10, 54–56, 57, 61, 183–185
Research Fund 7, 41, 44, 49, 52–56, 60–64, 70–71, 110–111, 122, 166
riots 24–34, 80, 123, 133, 139
Robinson, Robert 64–66, 83, 109–110, 121, 125
Roosevelt, F. D. R. 79, 81, 90
rum 13, 24, 81–82, 119, 125

Safeguarding of Industries Act (1921) 41
Science in Africa 51, 60
Scientific Advisory Committee 56, 64
Scientific Committee for Examining Alternative Uses of Colonial Raw Materials (SCEAURM) 42–43
scientists
  committees at the Colonial Office 55–62, 64, 157–158
  criticisms of 157–160, 162, 165–166
  independence of 7, 50, 59–61, 115, 119, 156, 167, 177–178
  status 8–9, 54, 58–64, 119, 123
Second World War 5, 11, 31–35, 37, 51, 58, 66, 106–108, 131, 164–165, 178
Shenfield, A. A. 130, 135–140, 149, 188

[205]

INDEX

Simonsen, John 43, 64–69, 83, 106–113, 119, 121–124, 157–158, 162, 171, 173
St Kitts 23, 24, 26, 110, 172
Stacey, Maurice 163, 176–177
Stanley, Oliver 64, 69, 77–78, 84, 86–87
Stockdale, Frank 25, 32, 34, 37, 39–44, 79, 84, 91, 110, 171, 177
sugar
  consumption 33
  oversupply 22–23, 25–26, 35–36, 42, 174, 177
  preference 25, 26, 35, 44
  price 22–26
  rationing 163
  research 41, 68, 93, 104, 109–119, 155, 162, 169–171, 176, 186
sugar beet 24
sugar cane wax 114–118, 146, 168–170, 172
sugar estates 27–28, 30–31, 132–134, 172
Sugar Foundation 116, 155, 170
Sugar Technology Laboratory 111–125, 154, 167–171, 176–178
sugar workers 26–29, 30, 111, 133
synthetics 41, 64–70

Taft, Robert 82
tariffs 23, 41, 63, 77, 81, 88, 89, 92, 126, 137, 140, 144, 183
Tate & Lyle 26, 31, 41, 117, 172–175
Taussig, Charles 79–81, 91, 94
tax relief 81, 99, 115, 142, 175, 187
Thaysen, A. C. 39, 120–125, 161, 168, 174–175
trade unions 27, 133–138

Trinidad and Tobago
  economic problems 132–135
  labour unrest 26, 28
  oil industry 14, 28, 30, 32, 40, 107–108, 132–134, 136, 149
  Pioneer Industries 136–143, 145–149
  political reform 130, 132, 147
  sugar industry 12, 22–26, 30, 32, 34, 132–133, 135
Trinidad Guardian 112, 119, 124, 134–135
Trinidad Leaseholds 28, 32, 69, 109, 133, 136
Tropical Products Institute 165
Tugwell, Rex 81–82

unemployment 1, 5–6, 27, 32–33, 75, 77, 80, 138, 148–149, 172, 189, 191
United Molasses Company 40, 106
United Nations 95
United States
  anti-imperialism 32, 34, 51
  domestic security 80, 90
  free trade 89–92
  wages 30

Weizmann, Chaim 38–39, 42–43
West India Royal Commission
  see Moyne Commission
West Indian Conference, Barbados (March 1944) 79–84
West Indies Federation
  see Federation
Wiggins, Leslie 65–66, 111, 114–123, 146, 169–170, 176–177
Williams, Eric 81, 95–96, 130, 140, 147–149, 190
Worthington, E. B. 50–60

Lightning Source UK Ltd.
Milton Keynes UK
UKHW010617230819
348396UK00003B/208/P